PRACTICAL ASTRONOMY

Observations, Experiments and Exercises

SEVENTH PRINTING

U.O. Herrmann
B.C. Thompson

Department of Physics
The University of Texas at Arlington

KENDALL/HUNT PUBLISHING COMPANY
2460 Kerper Boulevard P.O. Box 539 Dubuque, Iowa 52004-0539

Cover photo by U.O. Herrman.
Solar Eclipse, February 26, 1979.

ISBN 0–8403–2109–0

Printed in the United States of America

B 402109 01

10 9 8 7

Contents

Preface

This manual was developed to meet the needs of an introductory astronomy course for non-science majors. It contains exercises for outdoor observations, regular laboratory experiments, and classroom or homework exercises.

The manual is designed to be used in conjunction with one of the numerous non-technical astronomy texts. Many exercises, however, stand on their own and can be done to advantage by any serious student of astronomy. The large number of exercises and problems discussed should allow the instructor to select those which are most pertinent to her/his course, disregard some units altogether and use only parts of still other units.

For astronomy courses which are taught without an observational laboratory, the instructor may concentrate on those units which can be done at home or in a laboratory room, and vice versa. Many of the exercises are flexible enough to allow the instructor to expand and/or modify them to her/his own particular goals.

The over-all emphasis of the manual is to allow the non-technical student and the interested amateur astronomer to become familiar with the tools and the methods of astronomy. Some of these methods are explained in great detail, for example the Doppler shift, the H-R diagram, and the Color and Luminosity of stars, because they are of fundamental importance in astronomy. Many textbooks, unfortunately, treat these concepts much too lightly and in this context, the manual can be used as an adjunct to a textbook. The tools used in this manual will help an amateur astronomer to understand his own instruments better.

The student is asked to construct a number of different models. Usually, one is faced with tabulated data about the solar system, for example, or about the luminosities of stars. These data consist mostly of very large numbers which by themselves do not mean much. To construct a scale model will help the student to grasp and to visualize his or her position in our vast universe. Since the availability of commercial instruments may be very limited, some units of the manual use self-made ones. They are presented in a very simple form and can be made much more elaborate by the students (or the instructor) if it is so desired.

One should not expect that all students should do all of the units in this manual. There may be, however, a sufficient number of students who would like to do as much as they can. Therefore, students may be encouraged to perform these exercises also on a voluntary basis, or to extend some exercises over an extended period of time—exceeding the duration of a formalized course.

Finally, we would like to acknowledge the support, the encouragement, and the suggestions made by our colleagues, Drs. N.E. Brener, W.B. Self, and S.C. Sharma with thanks.

1

THE HISTORY OF ASTRONOMY

Unit 1.1

Astronomers and Their Achievements

OBJECTIVE

To relate some discoveries and developments in astronomy with the people who produced them.

INTRODUCTION

Since the appearance of what we now call "Science" on this planet, innumerable people have contributed to its growth. As a human endeavour, science is relatively very young, yet it has influenced our lives to an enormous degree. Astronomy is one of the oldest sciences and since its very beginning was intimately concerned with the question of man's place in the universe.

"Man is the Measure of all Things," Heraklitos said about two and a half thousand years ago. Everything seemed to revolve around man, and everything seemed to exist only for the benefit of man. About 2,000 years later, the Earth was recognized to be one planet among others, revolving around a central star, the Sun. Today we realize that our Sun is just an average star among uncounted others, situated in a huge system of stars, the Milky Way Galaxy. This, our home galaxy is a member of the "local cluster of galaxies", among uncounted other clusters of galaxies. We are not any more the center of everything, and we have started to ask if there are other intelligences "out there".

The people who have contributed to this grandiose view of the universe, however, are almost lost among the kings and queens and generals and football stars and stars and superstars of every shade, who crowd our limited memory. It seems, therefore, only just to ask you to remember a few more names, among the many hundreds who have made astronomy into the exciting science it is today.

PROCEDURE

On the following pages you will find four columns, Time, Place, Name, and Contribution. Some of these columns are filled with the respective information, some are left blank. Fill-in the blanks. Consult your textbook for historical information. It is not at all important to remember that Newton lived from 1642 to 1727. It **IS** important to remember that much of Newton's work rested on the labors of Kepler, of Galilei, Brahe, and Copernicus and all of them together labored hard to overthrow Ptolemy's views, etc., etc.

Unit 1.1 Name ____________________

Date due __________ Instructor ____________________ Section __________ Seat ______

Time	Place	Name	Contribution
about 600 BC	Ionian Greece, Miletus		First scientist known by name, predicted solar eclipse, universe was water, the Earth a floating disk on an infinite ocean; among many other things.
	Alexandria in Egypt	Aristarchos	First known proponent of a "helio-centric system", the "Copernicus of Antiquity".
276–196 BC	Alexandria	Eratosthenes	
	Alexandria		Size and distance determination of Sun and Moon, first accurate star maps, precession of equinoxes, 6 brightness classes (magnitudes) of stars.
about 120–160 AD	Alexandria		Used "Epicycles" to explain apparent planetary motions. His system of the solar system remained essentially unchanged for the next one and a half millenia.
			Introduced the "helio-centric" system in his famous work: "De Revolutionibus Orbium Coelestium."
1546–1601	Uraniborg, Denmark and Prague, Czech.		Known mostly by his first name, as e.g. (???)'s nova, made precise observations in his own observatory, proved that comets are not phenomena taking place between earth and moon. His observations, especially of Mars, provided the basis of the work done by the man who became his successor as imperial astronomer and whom he had hired as a young assistant.
1571–1630		Johannes Kepler	
1564–1642		Galileo Galilei	

Unit 1.1 Name ______________________________

Date due ______________ Instructor ______________________ Section ____________ Seat _____

Time	Place	Name	Contribution
1642–1727			Undoubtedly one of the greatest geniuses who ever lived. Contributed extensively and fundamentally to mathematics, many parts of physics and gave the "explanations" (physical reasons) for many previously observed astronomical phenomena.
1738–1822	England	Wm. Herschel	
1784–1846			Measured the first helio-centric parallax (61 Cygni)
1811–1877		Leverrier	
1877–1957 and	Princeton, USA		
1873–1940	Denmark	Ejnar Hertzsprung	
1889–1953			Proved in 1925 that there are other galaxies besides our own, classified galaxies, showed that all distant galaxies recede from each other.
1879–1955	Germany, Switzerland, USA.		Would have liked to have been a "citizen of this planet". The greatest physicist of this century. Alexander Pope wrote a poem about his great predecessor: "Nature and Nature's laws lay hid in night. God said: "Let Newton be!" and all was light." To which Bertrand Russel wrote a sequel, in jest: "The devil, shouted: "Ho! Let ____________ be!" restored the status quo."

2

NAKED EYE OBSERVATIONS OF CELESTIAL OBJECTS

Unit 2.1

The Star-Tracker

OBJECTIVE

To understand, assemble, and use the star-tracker to find your way across the sky.

INTRODUCTION

The star-tracker is a simple device which will help you find the major constellations and stars in the night sky for any given date of the year and for any given hour of the day. Do not yet try to assemble it, continue to read! We need to discuss some fundamental concepts first.

When you look at the sky, the stars seem to be affixed to a large hemisphere, the so-called **CELESTIAL SPHERE.** You, as the observer, appear to be at the center of this sphere, which seems to rotate very slowly from East to West. This apparent rotation is due to the Earth's rotation in the opposite direction, from West to East. The axis of rotation is the Earth's axis.

Imagine you are standing at the Earth's Northpole. Looking straight up along the extension of the Earth's axis, you look at the pole of rotation, the so-called **NORTH CELESTIAL POLE (NCP).** As seen from the Northpole of the Earth, all stars then rotate in complete circles around the NCP. These circles are parallel to Earth's equator as well as the **CELESTIAL EQUATOR** which is simply the extension of Earth's equatorial plane into the sky. At the Northpole, these star circles are also parallel to the horizon. The latitude of the Northpole is 90° N, which is the angle between the horizon and the NCP, (see Unit 5.4 on p. 5–18 for a proof of this statement). The north-star, Polaris, is **NOT** exactly at the NCP. It is about 0.8° off, and therefore describes also a circular path around the NCP. The time of one complete rotation of the celestial sphere is 1 sidereal day or 24 sidereal hours. One sidereal day is about 4 minutes shorter than one solar day, which is the basis of our civil time.

Now imagine you make a fast trip from the Northpole to the equator of the Earth, keeping the NCP and Polaris in sight. The farther south you go the more Polaris will seem to move towards the horizon. Suppose you make a break in your travel at latitude 33° N, drink a Coke, measure the angle between the horizon and the NCP, find it to be 33°, nod understandably, and continue on your way. Having arrived at the equator, you find the NCP at the horizon and Polaris very close to it. The stars would now **ALL** rise perpendicularly to the horizon, describe perfect semi-circles and set perpendicularly to the horizon. Looking straight up into the sky, you look towards the **CELESTIAL EQUATOR.** If you measure the angle between the horizon and the NCP you will find it to be 0°. The latitude of the equator is 0°.

Considering the apparent motion of the stars during one day:

AT THE NORTHPOLE you could see half of the stars all the time; all the stars you are able to see are **CIRCUMPOLAR.**

AT AN INTERMEDIATE LATITUDE you could see some stars all the time, i.e. they would always be **ABOVE** the horizon, the circumpolar stars. All other stars would rise, be above the horizon for varying times, then would set.

AT THE EQUATOR you would see all the stars, but only half the time. No star is circumpolar, all stars rise and set and are above the horizon for 12 hours. They reach their highest point above the horizon when they seem to move through a line which goes from the Northpoint on the horizon through the overhead point (the **ZENITH**) to the Southpoint on the horizon.

This line is called the **CELESTIAL MERIDIAN.** Consult Fig. 1 for the relationships between the terms discussed above.

In a planetarium, this apparent motion of the celestial sphere is beautifully reproduced since the star images are projected onto a hemisphere. All star maps—and your star-tracker as well—suffer from the fact that the apparent spherical view is reproduced on a plane sheet of paper. That results in some distortion. To minimize this effect, the star-tracker contains only stars and constellations adjacent to the celestial equator. The circumpolar stars are discussed in Unit 2.2.

The star-tracker consists of two parts, the star map forming the inner cylinder, and the horizon mask with windows, forming the outer cylinder.

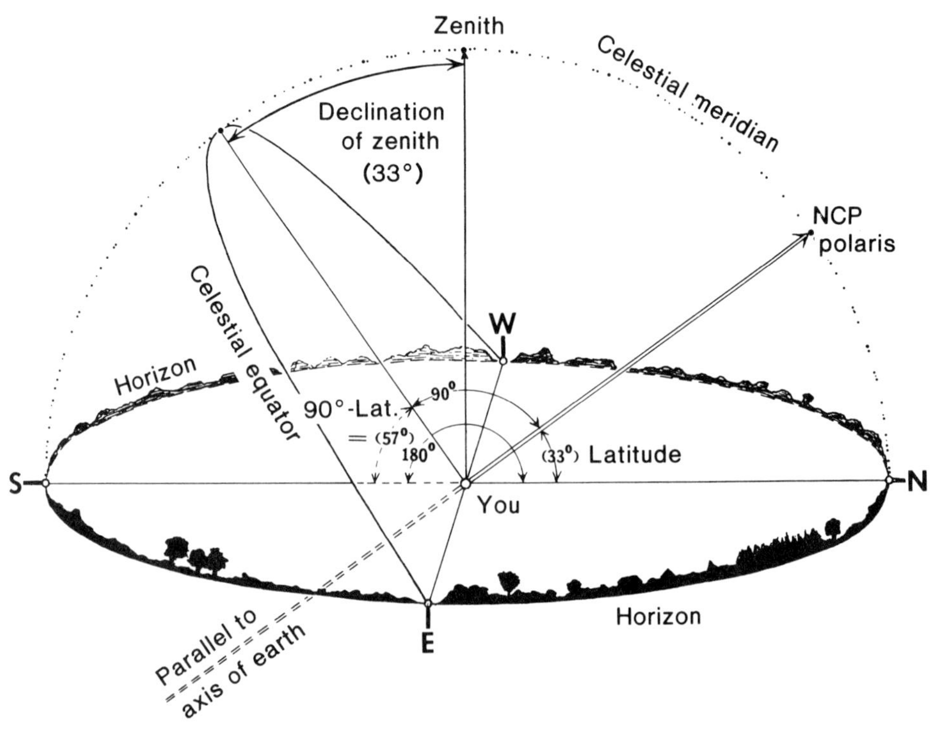

Figure 1

THE STAR MAP, P. 2–5 AND AT THE END OF THIS MANUAL

This map contains the brighter stars of the major constellations, connected by dashed lines which show the asterism of the constellation. Since the star-tracker is intended only as a first aid in finding your way across the sky, the constellations are plotted roughly to scale. All named stars, however, are accurately plotted. If you need to find the positions of fainter stars in a constellation or other, less conspicuous constellations, consult the star maps on p. 9–29 to 9–37.

The star map contains stars which lie within 60° of either side of the celestial equator. If we would use celestial coordinates, you would say: Stars with **DECLINATION** between +60° (North of cel. equator) and −60° (South of cel. equator). Which stars and constellations you see at any given time depends on your geographical latitude. **FIND OUT YOUR LATITUDE AND THEN DRAW A LINE PARALLEL TO THE CELESTIAL EQUATOR WITH DECLINATION VALUE EQUAL TO YOUR LATITUDE.** The Declination scale is on the left side of the star map. All stars and constellations on this line will be in your zenith at some time. This line will be called the Zenith-line, and you will see its value when the use of the star tracker is discussed.

At the bottom of the star map is the Right Ascension (RA) scale, expressed in hours. Remember that the celestial sphere rotates once completely around in 24 hours. This RA scale will be covered by the outer cylinder when the star-tracker is assembled. The same scale with the addition of degrees is repeated at the bottom of the page and will be readable through a window in the outer cylinder. Between these two RA scales is the DATE-scale, indicating the first and 15th day of each month, going from right to left. This datescale will also be readable through a window in the outer cylinder. Below the datescale is an hour scale, with each half and quarter hour indicated by tick marks, readable through the same window as the datescale after assembly.

On the star map, the following abbreviations are used for the constellations, listed from 0 hr RA and from top to bottom. Look at the star map while you read off the constellations and main stars on the following page.

THE HORIZON MASK, P. 2–7 AND P. 2–9

It depends on your geographical latitude which constellations and stars are visible at various times. The horizon mask will cover those constellations not visible at your location. If you live between latitudes 30° and 40° N, consult the horizon mask on p. 2–7, if you live between latitudes 40° N and 50° N, consult and use the horizon mask on p. 2–9.

The wide, curved line is your horizon. The major points of the compass are shown, i.e. N, NE, E, SE, S, SW, W, and NW. The azimuth angles are included for every 10° and numbered for every 20°. The lowest point of the horizon curve is the South-point of the horizon. The straight line from the S-point down is your meridian. The celestial meridian would be the continuation of this straight line upwards from the S-point—through the stars and constellations. When your star-tracker is assembled, the celestial meridian would then go from the N-point on the horizon straight up through the NCP (on the axis of your assembled star-tracker) and close to Polaris, through the overhead point (the Zenith) and down to the S-point on your horizon.

CONSTELLATIONS				MAIN STARS
abb.	astronomical name	common name	asterism	
CAS	Cassiopeia	—	W, M, 3 or $\sum$	β—Cas (not shown)
AND	Andromeda	—	—	α—And
ARI	Aries	ram	—	—
CET	Cetus	whale	—	—
PER	Perseus	—	—	—
TAU	Taurus	bull	V	Aldebaran
AUR	Auriga	charioteer	pentagon	Capella
ORI	Orion	—	—	Betelgeuze (Bet.) Rigel
GEM	Gemini	twins	—	Castor (Cast.) Pollux (Poll.)
CMI	Canis Minor	smaller dog	not shown	Procyon
CMA	Canis Major	larger dog	—	Sirius
LEO	Leo	lion	sickle or inverted?	Regulus
UMA	Ursa Major	larger bear	big dipper	Mizar/Alcor (not shown)
CRV	Corvus	raven	small square	—
VIR	Virgo	virgin	not shown	Spica
BOO	Bootes	herdsman	kite	Arcturus
LIB	Libra	scales	—	—
HER	Hercules	—	butterfly	—
SCO	Scorpius	scorpion	—	Antares
LYR	Lyra	harp	parallelogram	Vega
SGR	Sagittarius	archer	teapot	—
CYG	Cygnus	swan	northern cross	Deneb
AQL	Aquila	eagle	—	Altair
CAP	Capricornus	goat	—	—
PEG	Pegasus	flying horse	great square	—
PSA	Pisces Austrinus	southern fish	not shown	Fomalhaut

HOW TO ASSEMBLE YOUR STAR-TRACKER

Your star-tracker will have to take a lot of handling, therefore it should be sturdy. Use either the star map printed on cover stock at the end of this manual or use the star map on p. 2–5 and glue or tape on heavier cardboard, wooden or plastic cylinder, etc. Before you cut anything, be certain that you have drawn the required Zenith-line.

Cut along the left side as indicated, remove all the cross-hatched area to the points of the arrows. Carefully bend the map around into a cylinder so that the two Declination scales touch each other. Glue or tape together. This will be the inner cylinder of your star-tracker.

Next, take the page with the appropriate horizon mask for your geographical latitude, p. 2–7 or 2–9. Cut out the three windows as indicated. Cut along the curved horizon along the upper border as indicated. All cross-hatched areas

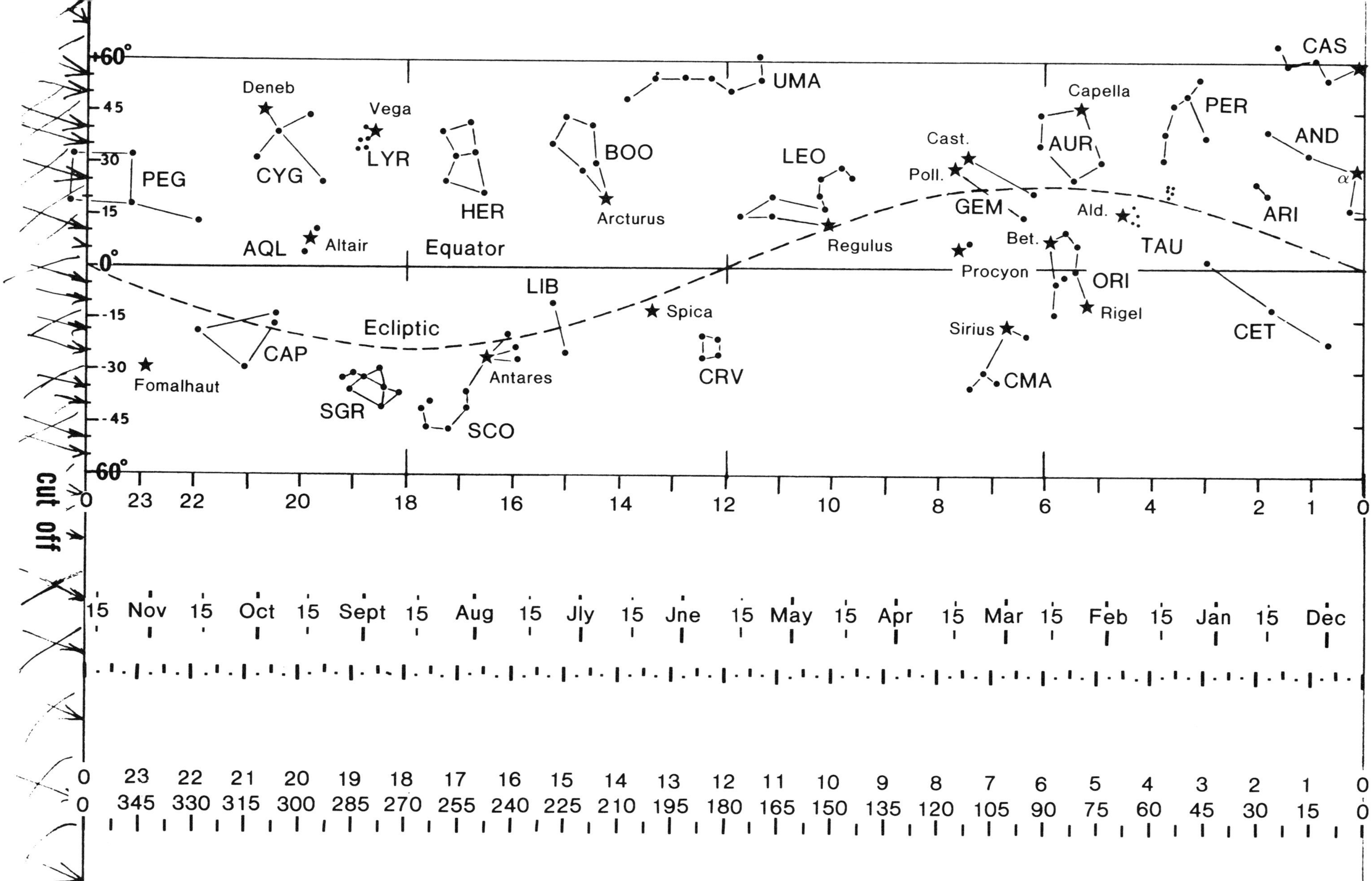
CAS
AND
ARI
PER
CET
TAU
Ald.
Capella
AUR
ORI
Rigel
Bet.
Procyon
CMA
Sirius
GEM
Cast.
Poll.
LEO
Regulus
UMA
CRV
Spica
BOO
Arcturus
LIB
Antares
SCO
HER
Equator
Ecliptic
SGR
Vega
LYR
Altair
AQL
Deneb
CYG
CAP
PEG
Fomalhaut
+60°
45
30
15
0°
-15
-30
-45
-60°
0 23 22 20 18 16 14 12 10 8 6 4 2 1 0
cut off
15 Nov 15 Oct 15 Sept 15 Aug 15 Jly 15 Jne 15 May 15 Apr 15 Mar 15 Feb 15 Jan 15 Dec
0 23 22 21 20 19 18 17 16 15 14 13 12 11 10 9 8 7 6 5 4 3 2 1 0
0 345 330 315 300 285 270 255 240 225 210 195 180 165 150 135 120 105 90 75 60 45 30 15 0

Figure 2

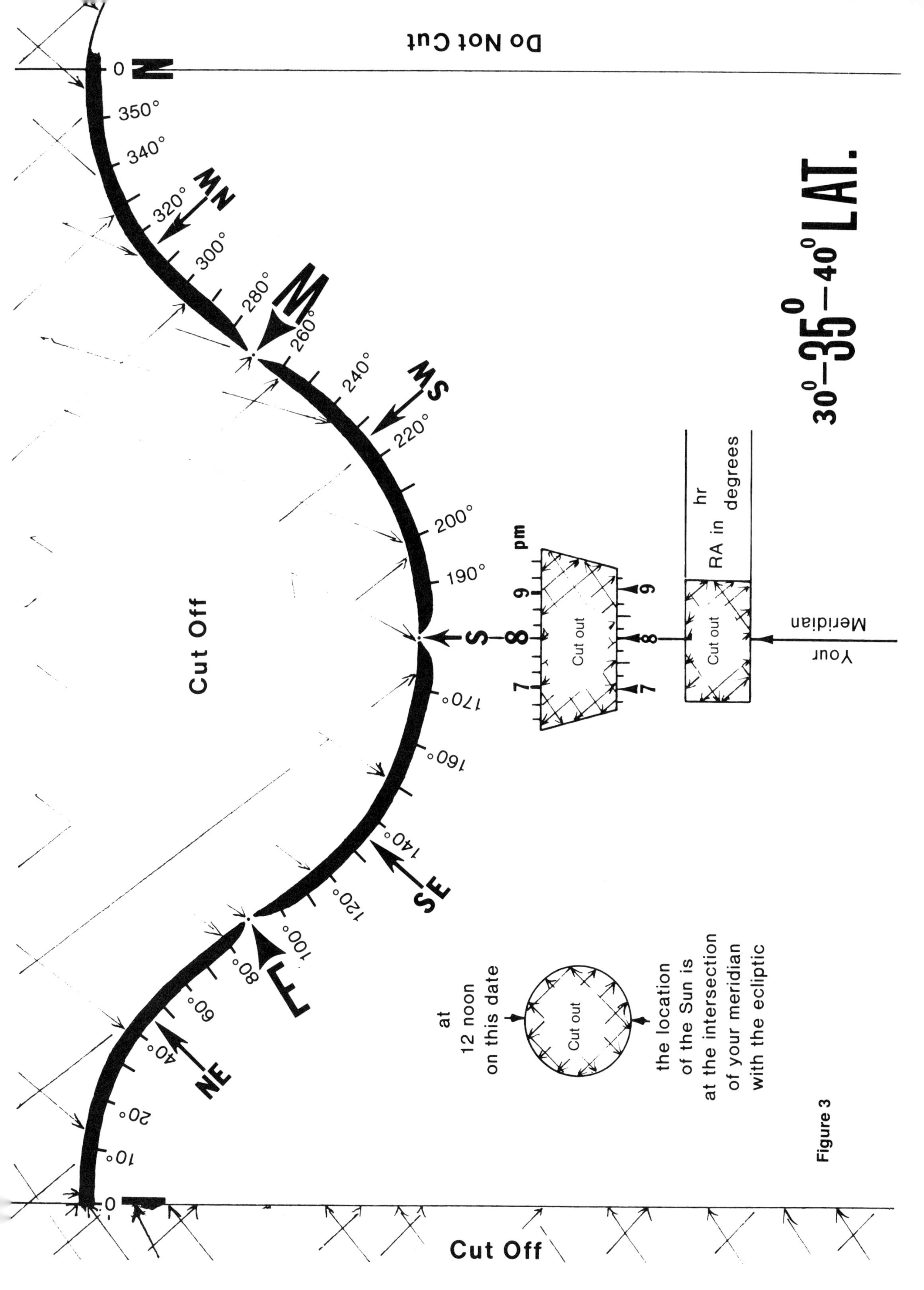

Do Not Cut
Cut Off
Cut Off
30°–35°–40° LAT.
N
NW
W
SW
S
SE
E
NE
0
10°
20°
40°
60°
80°
100°
120°
140°
160°
170°
190°
200°
220°
240°
260°
280°
300°
320°
340°
350°
0
pm
7
8
9
Cut out
Cut out
hr
RA in degrees
Your Meridian
at 12 noon on this date
Cut out
the location of the Sun is at the intersection of your meridian with the ecliptic

Figure 3

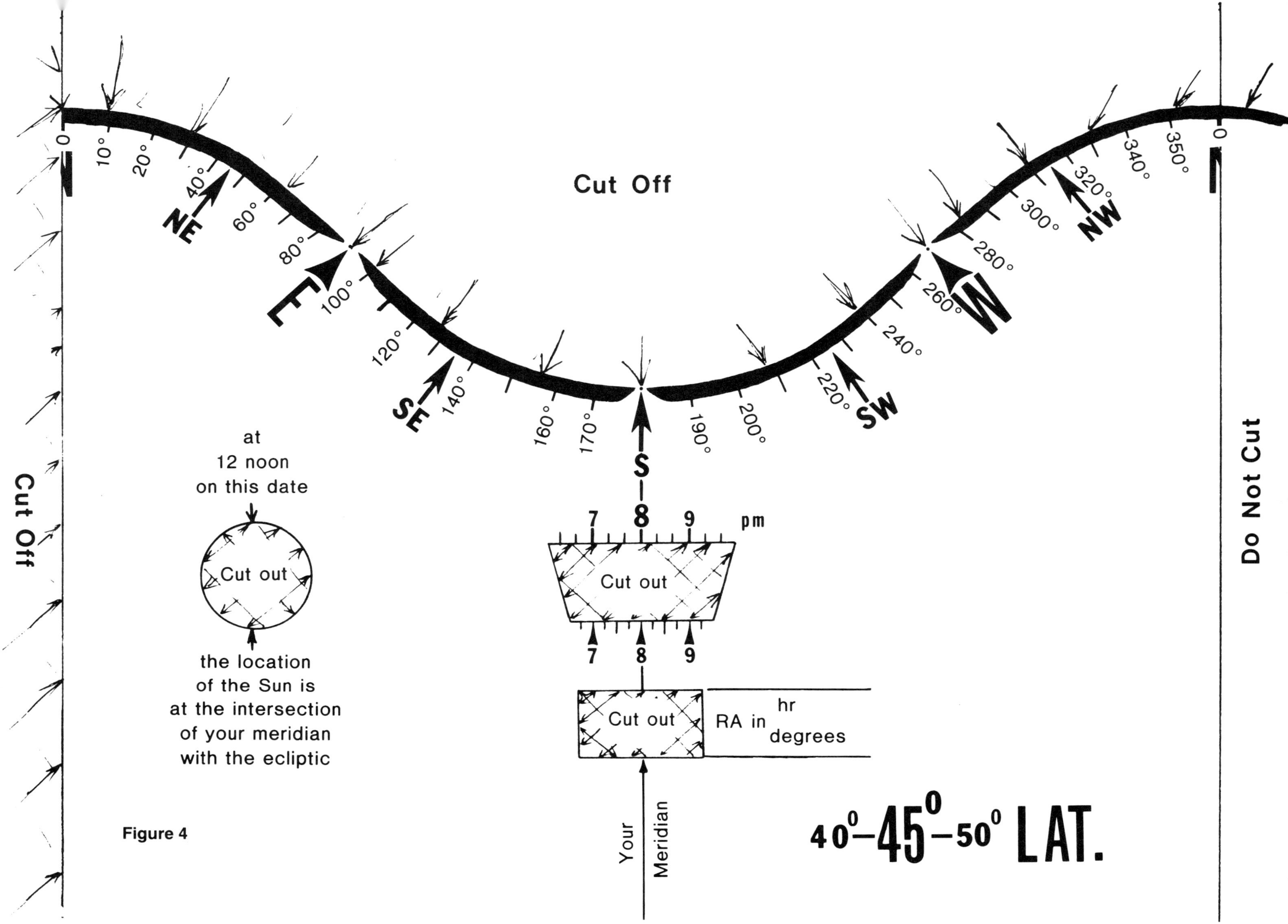
Cut Off
Cut Off
Do Not Cut
0
10°
20°
40°
60°
80°
100°
120°
140°
160°
170°
190°
200°
220°
240°
260°
280°
300°
320°
340°
350°
0
NE
E
SE
S
SW
W
NW
at
12 noon
on this date
Cut out
the location
of the Sun is
at the intersection
of your meridian
with the ecliptic
7 8 9 pm
Cut out
7 8 9
Cut out
RA in hr
degrees
Your
Meridian
40^0–45^0–50^0 LAT.

Figure 4

should be removed, including the left side. Now bend the horizon mask into a cylinder so that it fits over the inner cylinder. This outer cylinder should easily move over the inner cylinder. Glue or tape together. If the outer cylinder fits too loosely, you may insert some extra paper between the cylinder. If it fits too tightly, either cut it open again or use the spare horizon mask, p. 9–41 or p. 9–43.

HOW TO USE THE STAR TRACKER

Slip the horizon mask over the star cylinder and adjust **SO THAT THE E POINT WITH 90° AZIMUTH AND THE W POINT WITH 270° AZIMUTH ARE COINCIDENT WITH THE CELESTIAL EQUATOR ON THE STAR MAP.** Now rotate one or both of the cylinders until the desired date appears in the Date-window. All stars visible at your location will be above the horizon line on the date dialed at 8 pm (local) Standard Time.

EXAMPLE

Assume you live at latitude 33° N and you want to look at the stars on September 1 at 8 pm. You have used the horizon mask on p. 2–7. Looking directly south, the teapot of Sagittarius is going through your celestial meridian. It is about 20° above the horizon and as high in the sky as it will ever be at your location. Remember that it is 90° from the horizon to the zenith. You can therefore use the Declination scale for this measure, or you can make your own scale: 1 cm = 15°, or 1 mm = 1.5°. Whenever you want to measure the altitude of an object (angle above the horizon) you must measure vertically upward from the horizon.

Draw an imaginary line (or a thin, erasable pencil line) as the extension of your meridian line to the point where it intersects the Zenith-line. Your zenith—as you can now see—is just a little bit below (south of) the bright star Vega in the constellation Lyra the harp. Now look East. Tilt your cylinder until the horizon line is parallel to the actual horizon. You see the great square of Pegasus above the horizon. Looking towards the West, tilting your cylinder again until the horizon line is parallel to the actual horizon, you see Arcturus in Bootes about 30° above the Western horizon. Looking towards the NE, you see Cassiopeia.

Look again at the Date-window. Note that you will see the same part of the sky we discussed for September 1, 8 pm, also on September 15 at 7 pm, or as you could have seen on August 15, 9 pm, etc.

Below the Date scale is another scale, the Hour scale, visible through the same window. Put the lower 8 pm mark on a full hour tick mark of the Hour scale. You will see that the 7 pm mark as well as the 9 pm mark are also on a full hour tickmark. Each hour is divided into quarters. This hour scale can also be used to divide the 15 day time interval of the Date scale into finer parts. Note that 1 hour of time difference corresponds to about 15 days on the Date scale. The reason that it does not exactly coincide is the fact that the star map of 360° circumference (24 hr RA) has to be divided into 24 hours for the hour scale and into 365 days for the date scale.

Play around a little bit with your star-tracker until you have become familiar with these scales.

How will the sky look on the same day, September 1 at midnight? Set the date scale to September 1 for the 8 pm mark. Check that the E and W points coincide with the celestial equator. Note where the lower 8 pm mark points on the hour scale. Now move the outer cylinder 4 full hours (or 16 quarter hours). You can easily figure out in which direction: Stars rise somewhere in the East and set somewhere in the West. Looking at the date scale again after you moved the outer cylinder 4 hours you will note that the sky on September 1 at midnight looks just as the sky on November 1 at 8 pm. Auriga is above the horizon in the NE, the two handle stars of the teapot of Sagittarius are just still above the horizon and will be gone shortly, looking straight up to the zenith you see the great square of Pegasus. The big dipper is on the northern horizon.

Turn the cylinder back to September 1, 8 pm and look at the RA window. You read: 18 hr 40 min or about 280° RA. This reading is the RA for all stars which are on your celestial meridian. A star in Sagittarius has this RA and a Declination of about −30° (below the celestial equator). The bright star Vega in Lyra has about the same RA value while its Declination is about +40° (above the equator). You may thus read the RA angle of any object by moving the star map and reading at your meridian line.

HOW TO USE THE STAR-TRACKER TO LOCATE THE SUN'S POSITION

While you have seen the night sky on September 1 at 8 pm, the sun was, of course, below the horizon. Where? As you already know, somewhere on the ecliptic, the dashed line in the star map. Dial September 1 in the **CIRCULAR** window which is 8 hrs to the left of the 8 pm mark in the date-window. The sun's position in the sky is now a point on the ecliptic where your celestial meridian crosses the ecliptic. Take a ruler, extend your meridian line to the ecliptic and make a little pencil dot. In what Zodiac constellations is the sun? In Leo. Note that a constellation is larger than shown by its asterism lines. To repeat: On September 1, the sun's position is in the constellation Leo.

You may now use this pencil dot to find out where the sun rose in the morning of September 1. Keep the sun's dot in sight and rotate the outer cylinder until the dot appears just above the eastern horizon. You read: about 10° N of E or at azimuth of about 80°. Where on the horizon will the sun set on September 1? Rotate the horizon mask in the other direction until the sun's dot appears just above the western horizon. It sets about 10° N of W or at azimuth of about 280°. Try this for any other date of year, particularly for the equinoxes and the solstices.

You can do more with the sun's location but it requires more information than we are prepared to use at this time. You would need to know your geographical longitude as well as to be familiar with Standard Time and local solar time. For a discussion of time measures and some exercises see Unit 5.4 on p. 5–17. In addition to finding out where on the horizon the sun will rise and set, we ought to be able to tell at what times the sun will rise or set. Think about it.

EXERCISES

Use p. 2–13, to p. 2–17, as instructed. Fill in the blank spaces.

Unit 2.1 Name ______________________________

Date due ____________ Instructor ____________________ Section __________ Seat ____

You want to observe the sky on February 15 at 8 pm. Set your star-tracker to this date and time. Your meridian passes through which constellations? List them:

1) ______________ 2) ______________ 3) ______________

Which named star(s) is nearest your celestial meridian?

1) ______________ 2) ______________ 3) ______________

What is the RA of this star (round off to nearest 10 min or 5°)

1) RA = ___° = ___ hr ___ min 2) RA = ___ hr ___ min 3) RA = ___ hr ___ min

LIST ALL CONSTELLATIONS THAT YOU COULD SEE (EXCLUDING URSA MAJOR AND CASSIOPEIA):

(If a constellation is not fully visible, it is either rising or setting. Note.)

A) On Feb. 15 at 8 pm	B) Mark by (X) those constellations which will have set by midnight.	C) List all constellations which have appeared since 8 pm and are visible now at midnight.
1)		
2)		
3)		
4)		
5)		
6)		
7)		
8)		
9)		
10)		
11)		
12)		
13)		
14)		
15)		

Unit 2.1 Name ______________________________

Date due ____________ Instructor ____________________ Section __________ Seat ____

You want to observe the sky on ____________ at ______ pm. Set your star-tracker to this date and time. Your meridian passes through which constellations? List them:

1) ____________________ 2) ____________________ 3) ____________________

Which named star(s) is nearest your celestial meridian?

1) ____________________ 2) ____________________ 3) ____________________

What is the RA of this star (round off to nearest 10 min or 5°)

1) RA = ____° = ____ hr ____ min 2) RA = ____ hr ____ min 3) RA = ____ hr ____ min

LIST ALL CONSTELLATIONS THAT YOU COULD SEE
(EXCLUDING URSA MAJOR AND CASSIOPEIA):

(If a constellation is not fully visible, it is either rising or setting. Note.)

A) On ____________ at ____ pm	B) Mark by (X) those constellations which will have set by midnight.	C) List all constellations which have appeared since ____ pm and are visible now at midnight.
1)		
2)		
3)		
4)		
5)		
6)		
7)		
8)		
9)		
10)		
11)		
12)		
13)		
14)		
15)		

Unit 2.1 Name ______________________________

Date due ______________ Instructor ______________________ Section ____________ Seat ____

At what azimuth (nearest 5°) will the following stars a) rise and b) set? Express this azimuth angle in degrees and as N of E, or S of E, or S of W, or N of W.

Star	rises		sets	
	azimuth angle	°X of Y	azimuth angle	°X of Y
1) Regulus				
2) Aldebaran				
3) Capella				
4) Sirius				
5) Deneb				
6) Altair				
7) Antares				
8) Spica				
9) Arcturus				
10)				
11)				
12)				

Unit 2.2
Circumpolar Constellations

OBJECTIVE

To observe and to become familiar with the North-Circumpolar Constellations, to locate the Northstar, Polaris, and to determine geographic North.

INTRODUCTION

Among the 88 astronomical constellations there is one which almost anybody has seen: The **BIG DIPPER.** It is, however, not really a constellation but a so-called **ASTERISM.** The constellation Ursa Major, the larger bear, is much larger and contains the big dipper as its most conspicuous part. In general, an asterism is a simple picture achieved by connecting the brightest stars of a constellation by straight lines, and is usually part of a constellation. Asterisms are therefore much easier to recognize and are invaluable to help us find our way throughout the night-sky. Once you have recognized one asterism or one constellation in the sky, you can then find your way to other asterisms and constellations by consulting the appropriate star maps and finding adjacent constellations. For this reason, it is very convenient to start locating constellations by finding the Big Dipper first. However, if you live in southern latitudes **AND** in a metropolitan area, the Big Dipper may not be visible. It may be very low on the northern horizon and buried in all the pollution. Therefore, we need something else to start with. If the Big Dipper is not visible, then certainly Cassiopeia will be. Its asterism is called the "flying W" or "flying M". It will also be a northern circumpolar constellation. To begin with, we must

FIND THE NORTHERN HORIZON

The Sun sets somewhere in the West. If your left shoulder faces West, then you are looking North. Now consult any of the drawings in Fig. 1: You will see either the Big Dipper or Cassiopeia or both. As you can see from the drawings, their positions change; not only during the night but also during the year. The whole starry sky—the **CELESTIAL SPHERE**—seems to rotate once in 24 hours around a point which is very close to the North star, Polaris. This point is called the North Celestial Pole (NCP). Polaris is only 0.8° off. We will discuss the NCP and Polaris later, for now let us observe the northern sky a little longer.

THE NORTH CIRCUMPOLAR CONSTELLATIONS

Choose the drawing for the appropriate season from Fig. 1. This will give you a rough idea where the Big Dipper and Cassiopeia are located around 8 pm. If you look at any other time of the day, Fig. 1e shows the hourly change of these two constellations, plotted for 2 hour intervals for September 1.

A more accurate drawing is Fig. 2. Turn the drawing until your date of observation appears at the bottom. If it is 8 pm Standard Time, then both the Big Dipper and Cassiopeia are properly aligned, meaning that they appear in the sky as they do in Fig. 2. For example: Assume it is March 1, 8 pm Standard Time. Turn Fig. 2 until March 1 is at the bottom of the horizon circle. You see Cassiopeia as a reversed 3 or a sigma ($\sum$) high up in the Northwest. The Big Dipper stands on its handle and is high up in the Northeast. As you can see from Fig. 2 also, the positions of both constellations will change in the following manner: 2 hours later

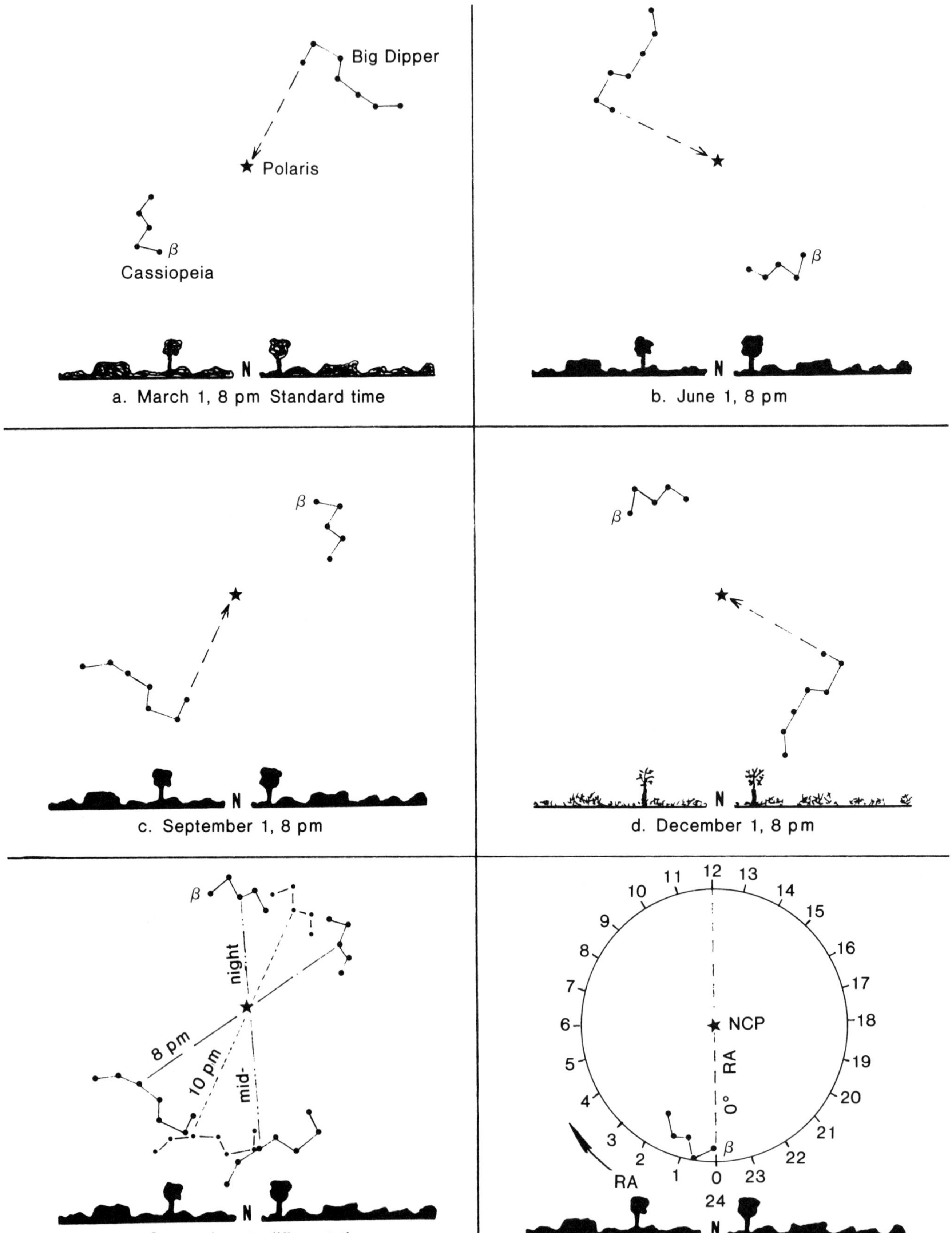

a. March 1, 8 pm Standard time

b. June 1, 8 pm

c. September 1, 8 pm

d. December 1, 8 pm

e. September 1, different times

f. Location of vernal equinox

Figure 1

(10 pm) each constellation has moved through an angle of 30° counterclockwise, or: 2 hours later, at 10 pm, both constellations appear as they would at 8 pm on April 1. Roughly speaking, "2 hours later" corresponds to "1 month later".

We will assume that you have identified both the Big Dipper and Cassiopeia. Where is the North star, Polaris? There are two easy ways to locate it:

a) Look for the pointer stars in the bowl of the Big Dipper, extend the distance between them about 5 times along the dashed line in Fig. 2 and you get Polaris.

b) If you can see both the Big Dipper and Cassiopeia, Polaris is about halfway between them. It is the only bright star in that part of the sky. Polaris is the main star of the constellation Ursa Minor with the asterism little dipper. Unless you live in the country and still enjoy fairly clear skies you will probably not see the little Dipper. Polaris is the only star of that constellation, bright enough to be easily seen.

FINDING NORTH

Having located Polaris, we now can determine the North point on the horizon. Drop an imaginary line straight down from Polaris. Where it touches the horizon is North—or 0° azimuth as you will later use it—however, only approximately. As stated earlier, Polaris is 0.8° off the NCP. Off, where?

Consult Fig. 2: At the center of the horizon circle is the NCP, denoted by +. If you draw a straight line from ϵ-Cassiopeiae to the NCP, the line will go through Polaris. It is useful to remember that ϵ-Cassiopeiae and Polaris are on the same side of the NCP. On the other hand, if you draw a line from Mizar/Alcor in the Big Dipper to the NCP, then you encounter the NCP first and beyond it you find Polaris. Amateur astronomers who want to set up their scopes accurately should make sure that the polar axis points at the NCP and not Polaris, particularly for photographic purposes.

NCP, CELESTIAL MERIDIAN, CELESTIAL EQUATOR, AND GEOGRAPHICAL LATITUDE

The North Celestial Pole (NCP) is defined as the point where the extension of the Earth's axis intersects the celestial sphere.

The line, starting at the North point on the horizon, going straight up, through the NCP (close to Polaris), through the overhead point, the Zenith, and ending at the South point on the horizon is called the **CELESTIAL MERIDIAN.** Imagine an Earth globe. The axis goes through the N and S poles. The equator is a circle at right angles to the axis. Now look North. Extend one arm towards the NCP (which is parallel to the axis of the Earth). Point your other arm at a right angle to the first one towards the East. Move this arm from East towards the West, always keeping it at right angles to the arm pointing at the NCP. Your arm traces the **CELESTIAL EQUATOR** in the sky. Remember that the celestial equator is always at right angles to the direction to the NCP.

Now stretch one arm out towards the North point on the horizon. This would indicate an angle of 0° altitude. Move your arm up to the NCP. This angle between the horizon and the NCP (approximately Polaris) is your geographical latitude. For a proof of this statement see Fig. 2 in Unit 5.4, p. 5–19. To make the same statement in other words: The altitude of the NCP is your geographical latitude.

CASSIOPEIA AND THE VERNAL EQUINOX

Later in the course and in this manual we will discuss astronomical coordinates, specifically the so-called Right Ascension (RA). It is counted from the vernal (or spring) equinox with RA = 0. Many students are interested to "see" the vernal equinox in the sky. There is nothing to see, it is merely the point where the ecliptic (the apparent path of the sun through the Zodiac constellations) intersects the celestial equator on March 21 (the beginning of spring).

It so happens that the star β-in Cassiopeia has a RA of about 0. Look at Fig. 1f. The line from the NCP through β-Cassiopeiae is the 0 RA line. Where this line intersects the celestial equator is the vernal equinox. Use your arms to indicate the celestial equator as described above. Wherever β-Cassiopeiae is, extend the line from the NCP through it to the celestial equator.

EXERCISES

1) Take the map of circumpolar constellations (p. 2–31) with you for observation. Rotate it until the constellations in the map are aligned with the constellations in the sky. Identify the circumpolar constellations. Then consult, if necessary, the other star maps (pp. 2–33 to 2–39) and answer the questions of part 1 on p. 2–25.
2) Use either the starfinder for the Altitude/Azimuth set-up or a protractor for altitude measurement (Unit 5.3, p. 5–13). Measure the angle between the horizon and Polaris. Note in part 2 on p. 2–25.
3) Be certain that you know and can point out to your instructor or laboratory assistant:
   a) the NCP
   b) the celestial meridian
   c) the celestial equator
   d) the ecliptic
   e) the position (or location) of a constellation a specified time earlier or later in the same night.

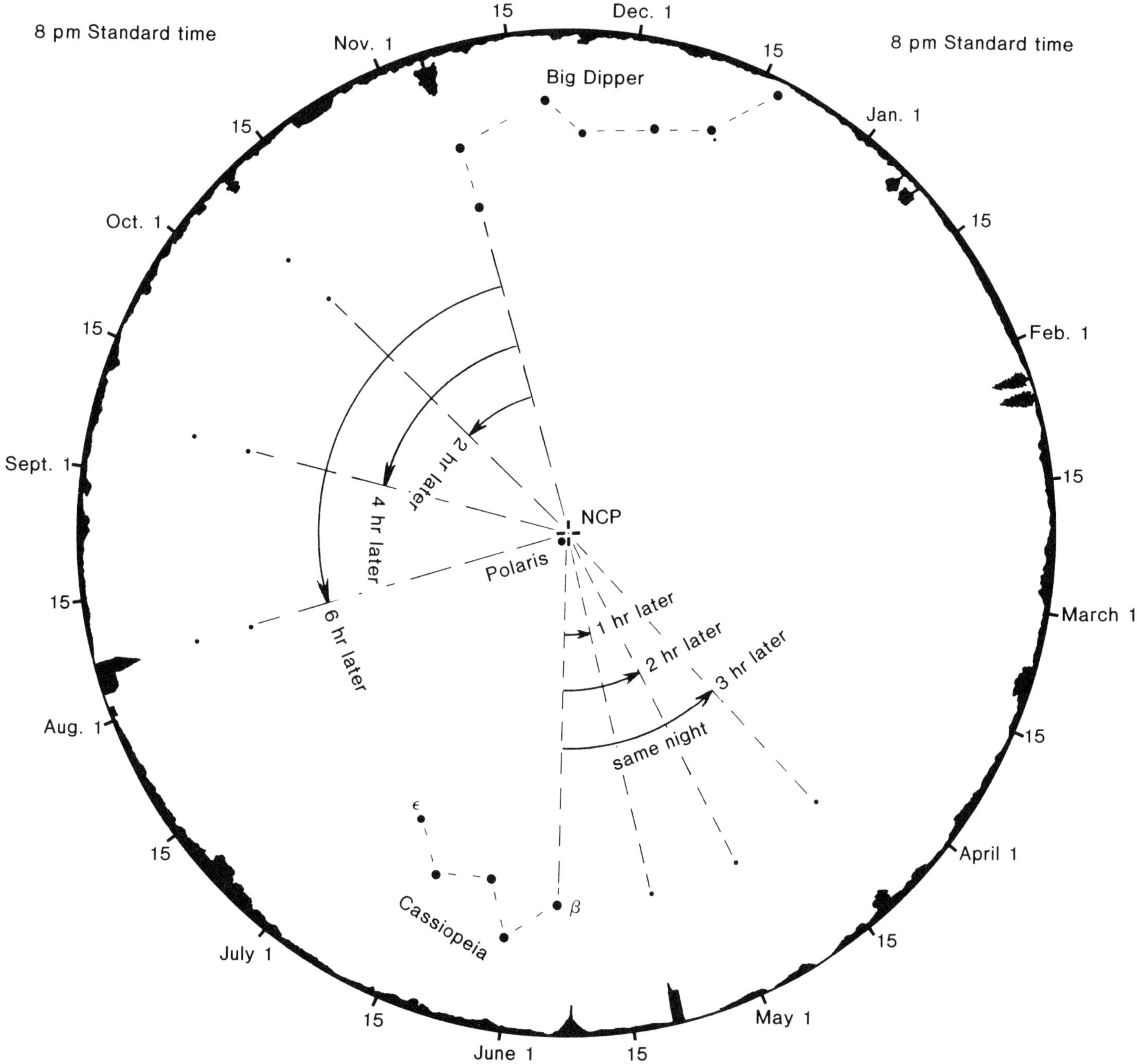
8 pm Standard time
8 pm Standard time
8 pm Standard time
8 pm Standard time
Dec. 1
15
Nov. 1
15
Jan. 1
15
Oct. 1
15
Feb. 1
15
Sept. 1
15
March 1
15
Aug. 1
15
April 1
15
July 1
15
May 1
15
June 1
15
Big Dipper
NCP
Polaris
2 hr later
4 hr later
6 hr later
1 hr later
2 hr later
3 hr later
same night
Cassiopeia
ϵ
β

Figure 2

Unit 2.2 Name ______________________________

Date due ____________ Instructor ____________________ Section ____________ Seat ______

1.1 Which constellations (asterisms) are circumpolar (always above your horizon) at your geographical location. If a constellation is partly circumpolar, note.

1) ____________________ 2) ____________________

3) ____________________ 4) ____________________

5) ____________________ 6) ____________________

7) ____________________ 8) ____________________

1.2 After how many hours would Cassiopeia and the Big Dipper have interchanged positions? ______ hr

1.3 Look at the drawings a) through e) of Fig. 1. Suppose you want to determine the north direction as accurately as possible by dropping a line (imaginary) from Polaris to the horizon. Which of the Cassiopeia-Polaris alignments would you choose? Specify drawing number, date, and time: ________

2.1 The angle between the horizon and Polaris was measured with what instrument:

____

2.2 The measured angle is: ____________ degrees = your geographical latitude.

2.3 The actual geographical latitude is ________ degrees N, as determined from: (List the source): ____

____

Unit 2.3
The Use of the Star Maps

OBJECTIVE

To learn to use the star maps to identify objects in the night sky and measure angular distances.

INTRODUCTION

Star maps or charts are nothing more than a representation of what one would see in a given part of the sky. There are five different maps, representing the total sky seen in the Northern hemisphere, used with this manual. Not all of the stars that are visible are plotted on these maps. The stars used on these maps represent approximately what one could see from an urban location. How bright a star appears to an observer on the surface of the Earth is called the apparent magnitude. The stars are represented by black dots on your maps with the size of the dot representing how bright the object will appear. The larger the dot, the brighter the object will appear. Note that in the apparent magnitude scale a smaller number represents a brighter star. The size of the dots and the corresponding magnitude numbers are printed on the map of circumpolar constellations. These maps are the equivalent of the maps that one has of the surface of the Earth. Any point on the Earth can be located by knowing the longitude and latitude of that point. (These coordinates are treated in detail in Unit 5.4). The most common coordinates used for locating points in the sky are Right Ascension and Declination. (These coordinates are treated in detail in Unit 5.5). The lines of declination are drawn around the celestial sphere parallel to the celestial equator and lines of Right Ascension are drawn between the celestial poles.

For the first use of the star maps one need not be concerned with the coordinates, but should learn how to use the maps to find what is visible at a given time. The star maps in this manual are divided into two types. The circular map represents that part of the sky that contains the constellations that would be circumpolar when observed from the Southern United States. This map will be used only for that part of the sky observed while facing North. Because the asterisms of the constellations are the most useful aids in orienting one relative to the sky each observer should know the asterisms visible at his location on Earth. The most common asterism in the Northern hemisphere is the "Big Dipper" (Ursa Major). To use the circular star map it is only necessary to rotate the map until the drawing of the "Big Dipper" is oriented as you see it in the sky. When you have done this the rest of the constellations on this map will be in the same relative position as the objects in the sky (see Unit 2.2).

The rectangular maps are somewhat more difficult to use. The four different rectangular maps represent the part of the sky that would pass your meridian in six hours. Thus the four maps represent the total sky (24 hours). To use these maps it is necessary to know that these maps are a projection of a spherically appearing sky on flat paper. Hence what you see in the sky will not have the normal orientation. These maps are drawn on a square grid with 2 mm = 1 degree, and North at the top of the page. For your maps to have the same orientation as the sky it is necessary to learn a few rules and to remember that the top of the maps must always point toward North, i.e. the star Polaris.

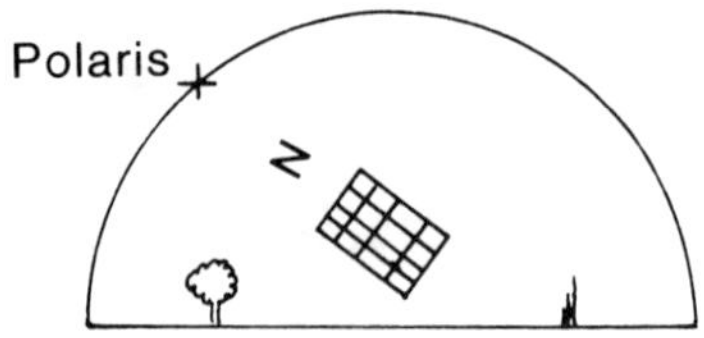

Figure 1

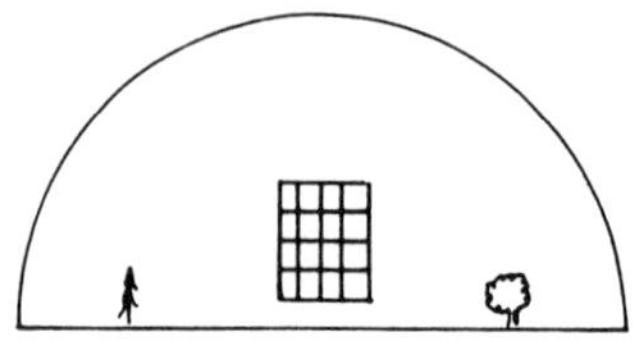

Figure 2

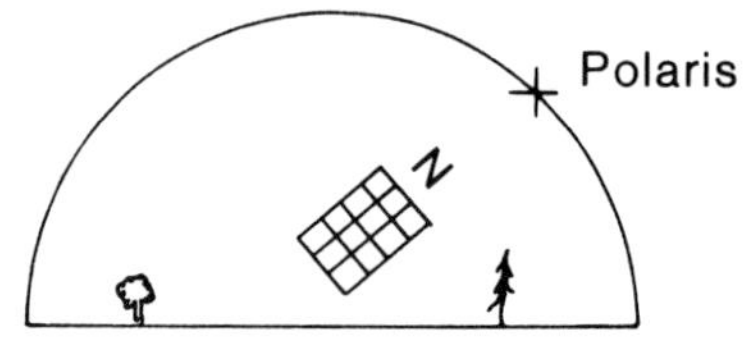

Figure 3

First, if you are facing East, one of the four rectangular maps will be used, the map containing the visible constellations for that time of year must be tilted as shown in Fig. 1.

If you are facing South the maps may be held as you normally hold a regular sheet of paper. This is shown in Fig. 2.

If you are facing West the top of the page must still point North as shown in Fig. 3.

By changing the orientation of your maps as described, the drawings on the maps will have the same orientation as the asterisms in the sky.

The rectangular maps are constructed such that approximate angular distances may also be measured from point to point on these maps. To determine the angular separation of any two points on these maps one needs only to transfer the distances to a piece of scrap paper and lay the paper along the scale at the bottom of the map to measure the angle.

EXAMPLE

Consider the two points, R.A. = 75°, Declination = +10°, (a point near ORION) and the point R.A. = 60°, Declination +20°, (a point between Aldebaran and the Pleiades). If you measure this distance point to point it is 36 mm. By using two marks on a sheet of paper and laying the paper along the scale at the bottom of the sheet you see that this distance is 18°. (Note that you could also have used the ruler measurement, 36 mm with 1 degree per 2 mm = 18°).

Unit 2.3 Name ________________________________

Date due ______________ Instructor ______________________ Section __________ Seat ____

By tilting the maps as shown in Fig. 1 and Fig. 3 you can determine the orientation of the asterism as a particular constellation rises and sets. Simply tilt the map so that the top points to Polaris and hold the sheet near the edge of a table. The table becomes the horizon and the orientation may be seen as it would appear in the sky.

1) Leo Rising Horizon	2) Leo Setting

3) Bootes Rising Horizon	4) Bootes Setting

5) The angle between Betelgeuze and Rigel is ________ degrees.

6) The angle between Aldebaran and the Pleiades is ________ degrees.

7) The angle between Castor and Pollux is ________ degrees.

8) The angle between Procyon and Sirius is ________ degrees.

9) The angle between the point R.A. = 150°, Dec = −10° and the point R.A. = 160° and Dec = 0° is ________ degrees.

10) The angle between the point R.A. = 75°, Dec = + 10° and the point R.A. = 95°, Dec = +5° is ________ degrees.

Unit 2.3 Name ______________________________

Date due ____________ Instructor ______________________ Section ____________ Seat _____

Stellar Magnitudes (apparent); all stars brighter than m = 3 are drawn in

- ● brighter than m = 1.0
- • fainter than m = 1.1 and brighter than m = 2.5
- • fainter than m = 2.6 and brighter than m = 4.0
- • fainter than m = 4.0

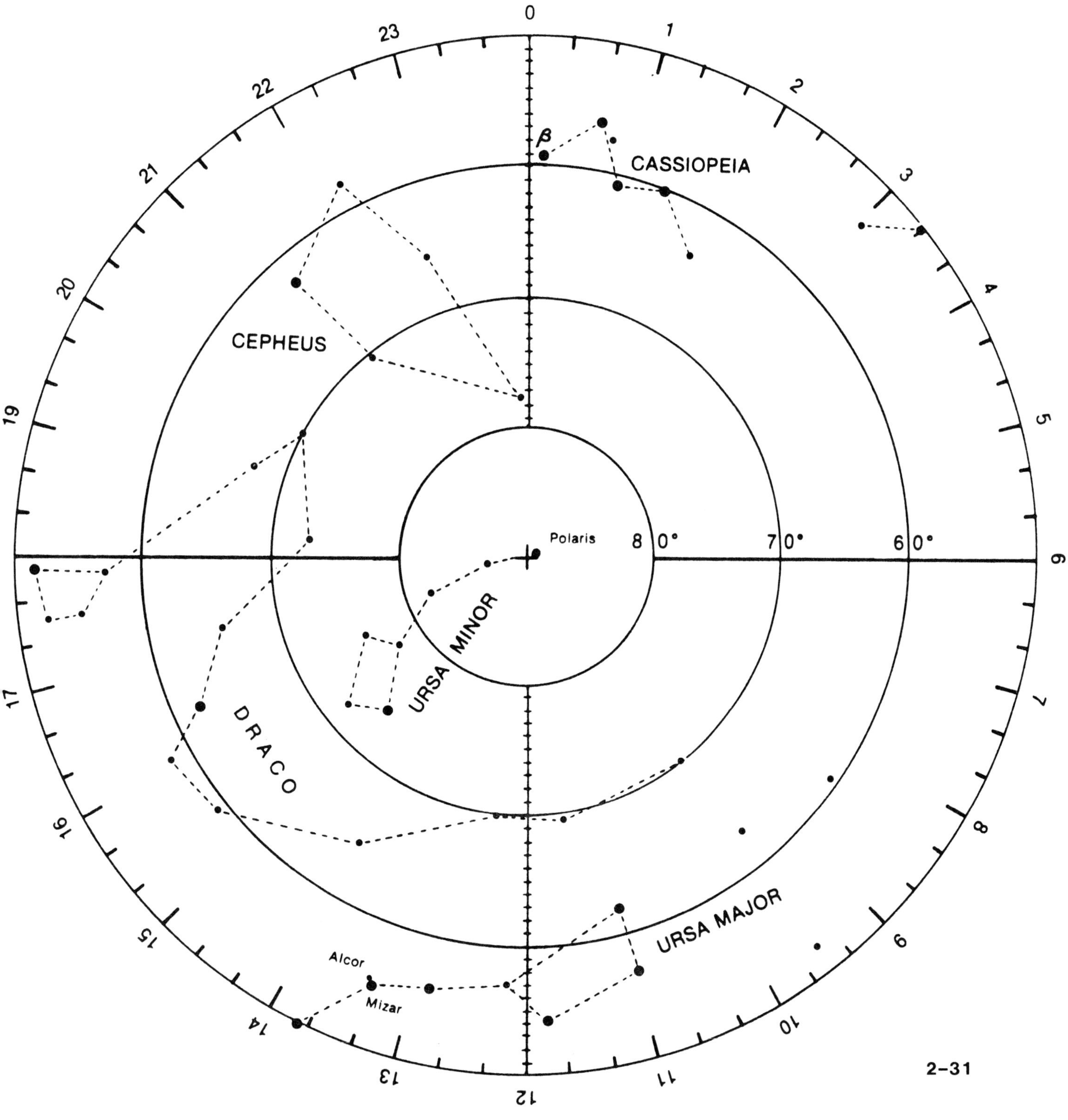

Unit 2.3 Name ____________________

Date due __________ Instructor ____________________ Section __________ Seat ______

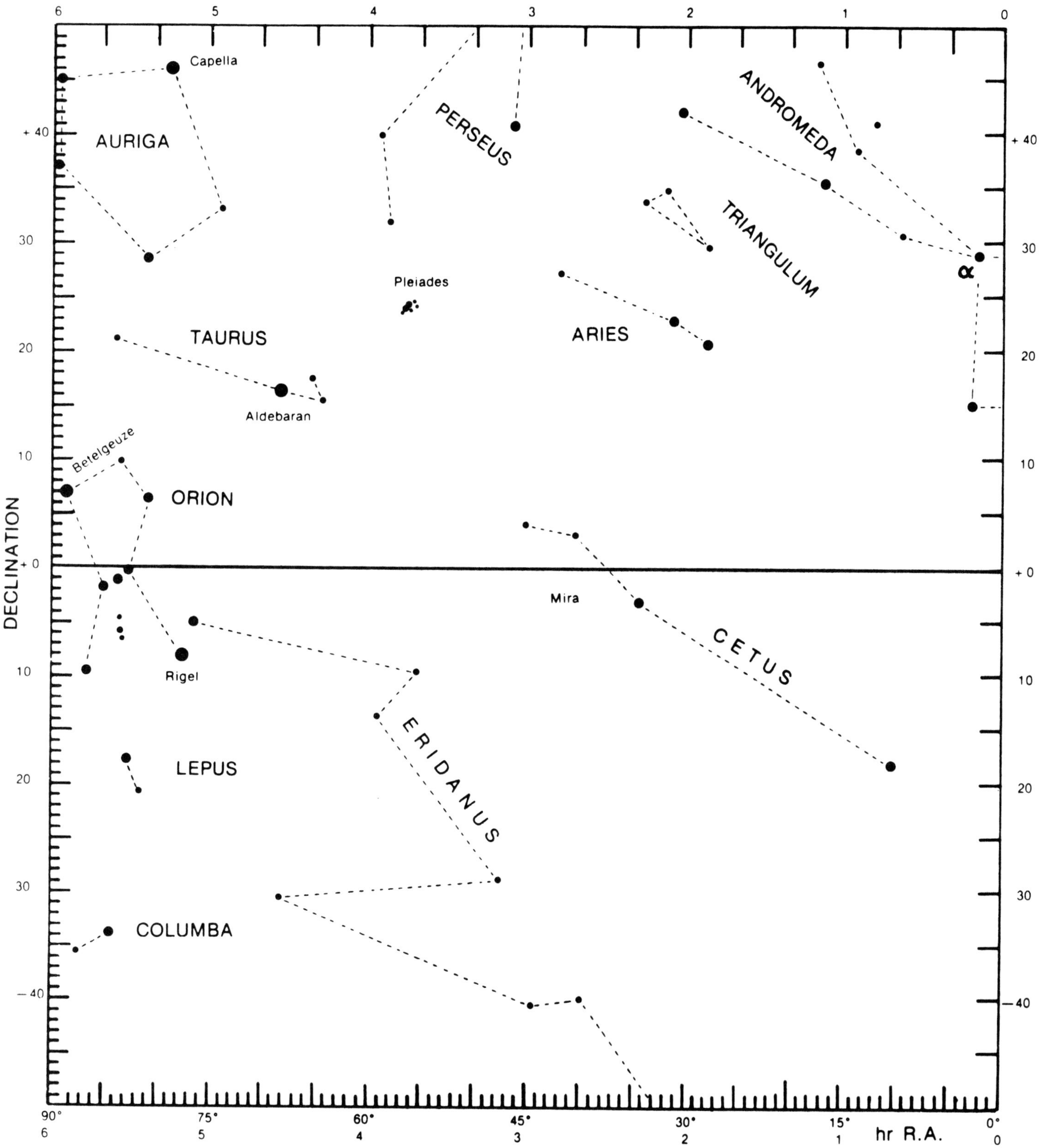

Unit 2.3 Name ______________________

Date due ____________ Instructor ______________________ Section ____________ Seat ______

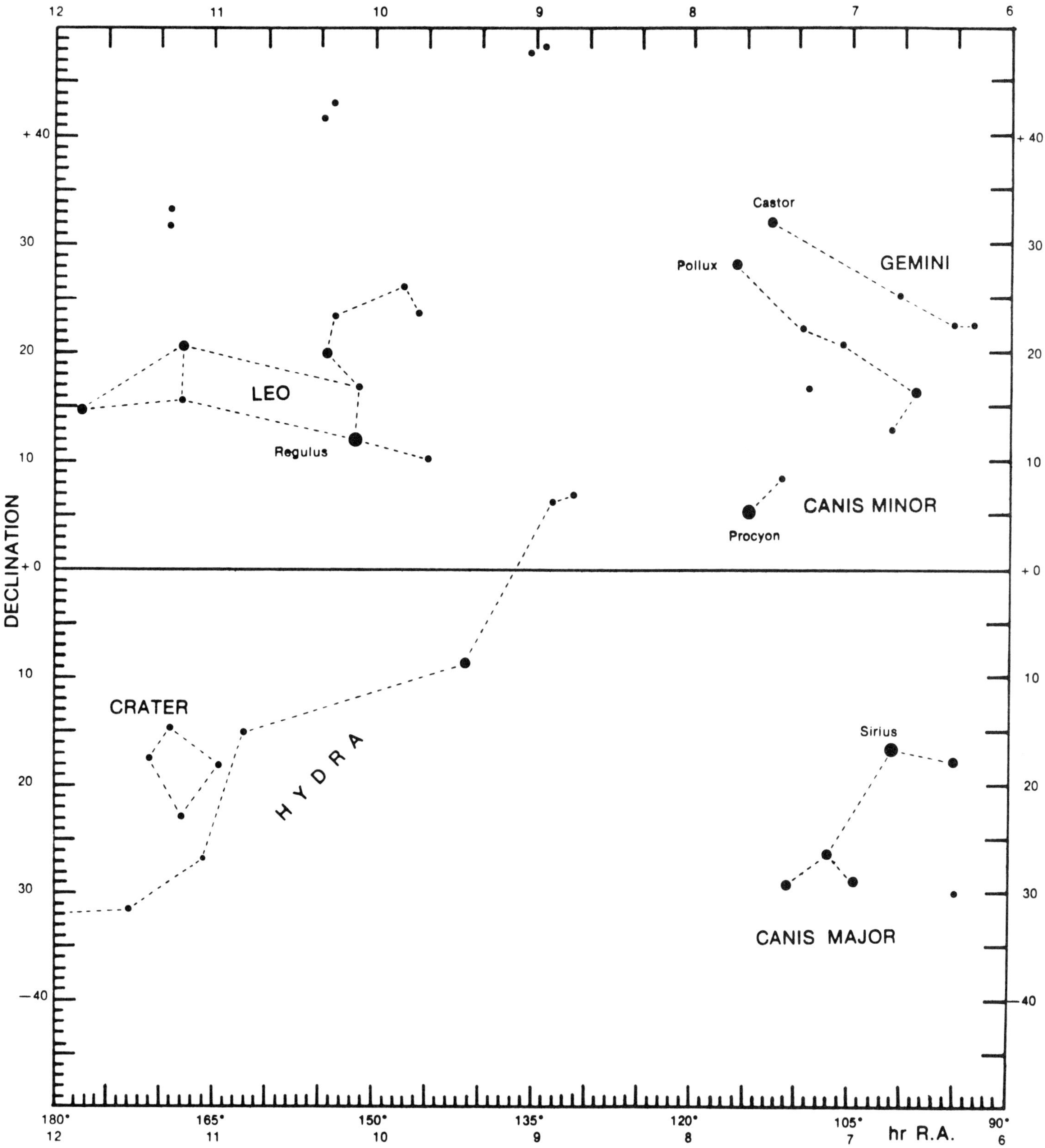

Unit 2.3 Name ______________________________

Date due ______________ Instructor ______________________ Section ____________ Seat ______

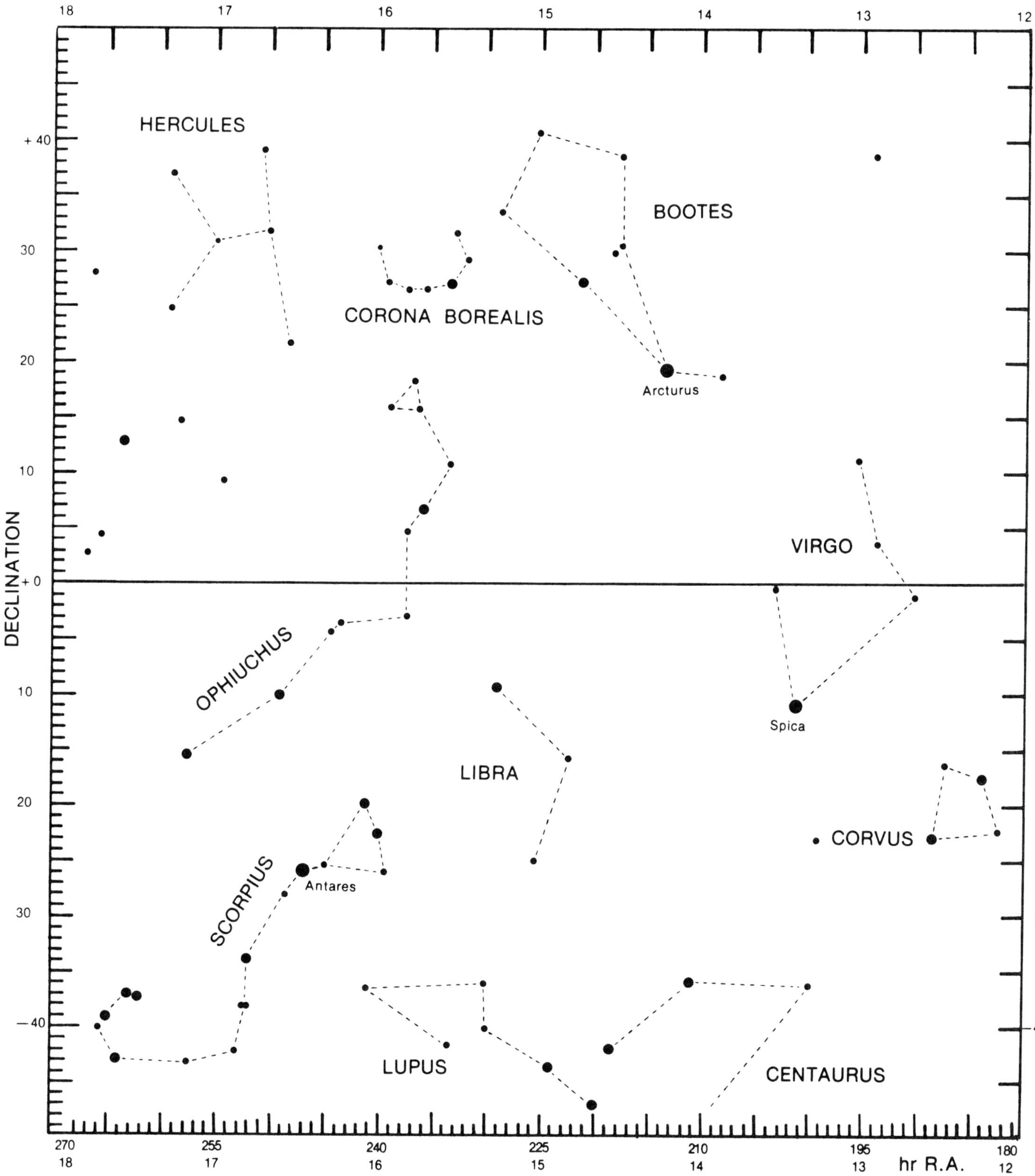

Unit 2.3 Name ______________________________

Date due ______________ Instructor ______________________ Section ____________ Seat ______

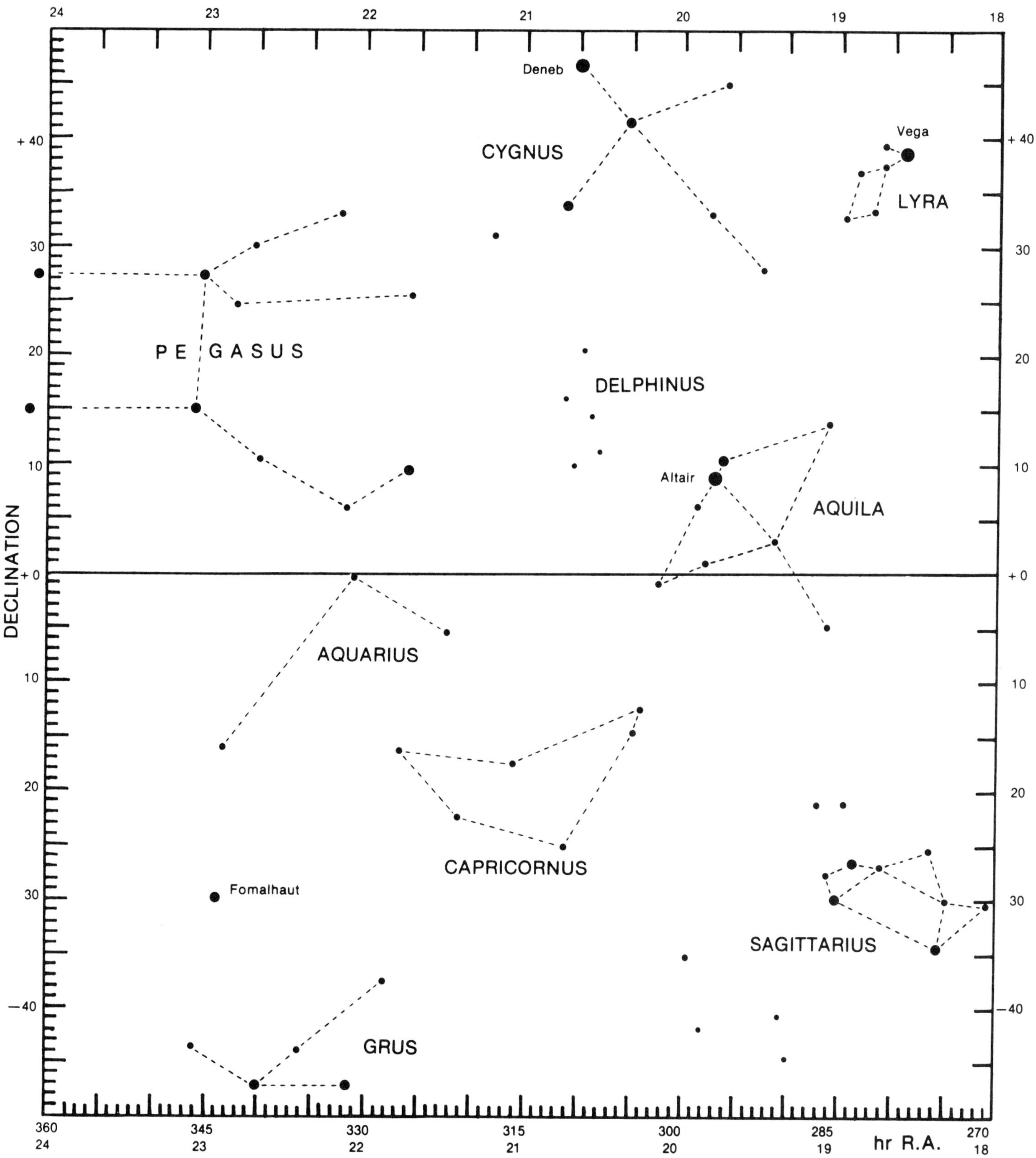

Unit 2.4

Observational Laboratory Requirements

INTRODUCTION

A Test will be given towards the end of the observation periods as specified by your instructor. For this test you need to know the information described below.

To prepare for the test, study the requirements, use your star-tracker, consult the star maps and attend the point-outs of constellations and stars by our laboratory assistants. You are welcome to listen to the pointouts as often as you like or feel necessary. The test, however, will be given only once per student in the semester. There is **ONE, SINGLE TEST.** You may take it on any scheduled clear night when at least 8 of the required constellations are visible.

When you feel ready to take the test, do the following:

a) Show your ID card to a lab assistant and let her/him take your name for the test.
b) Follow instructions as to the time and location for the lab test. Usually, a lab

   test is given every ________________ hour on a scheduled night.

   Tests are given between the dates of: ____________ and ____________ .

   ____________ ____________

   ____________ ____________

REQUIREMENTS

The requirements are listed in the box below according to sections. Written across the box are the requirements common for both planetary and stellar astronomy, or for course/section ________________ and course/section ____________ . Below each section are further requirements for each course/section separately.

YOU MUST KNOW

Planetary Astronomy	Stellar Astronomy
course: ______ section: ______	course: ______ section: ______

1) All constellations by their astronomical and common name which are displayed on your star-tracker and which are visible at the date and time you take the test.
2) The names and locations of all named stars within a constellation on the star-tracker, visible as specified under 1).
3) The names and/or the astronomical designations of the following Messier objects which are located within the visible constellations as specified under 1).

   Messier objects: ________________________________

4) The locus of the celestial meridian
5) The locus of the celestial equator
6) The locus of the ecliptic
7) The position of a certain star or constellation a certain time earlier or later than the test time.

Planetary Astronomy	Stellar Astronomy
8) The definition of the North Celestial Pole	8) The color and M-K class of the named stars on the star-tracker
9) The relation of Polaris to the NCP	9) The physical reason for the observed difference in the color of stars.
10) The significance of the angle between the horizon and the NCP	10) Additional requirements as specified by your instructor
11) Additional requirements as specified by your instructor	
________________________	________________________
________________________	________________________
________________________	________________________
________________________	________________________
________________________	________________________
________________________	________________________
________________________	________________________
________________________	________________________

3

TOOLS FOR SIMPLE ASTRONOMICAL MEASUREMENTS

Unit 3.1

Measuring Angles with a Protractor

OBJECTIVE

To make your own simple measuring device and use it for some measurements.

INTRODUCTION

In many of the units of this manual you are asked to measure angles. You will use the starfinder for some of these measurements because it is very convenient whenever you have to measure two angles which have a certain relation to each other, for example Right Ascension and Declination. For other measurements a simple protractor with a sighting device and possibly a plumb line is quite sufficient and much easier to use.

APPARATUS

1 protractor, either full or half circle with a minimum diameter of 4 inches.
1 straw or any other sighting device.
1 thin rod (not more than 3/16 of an inch in diameter).
A piece of thin string and a weight (a small rock will do).
A stand if necessary.

PROCEDURE

Instead of giving you detailed instructions about the assembly of this device, do it yourself. Chances are that you will come up with a better device than instructions could provide. Nevertheless, look at the pictures below. Each of them shows the application and construction for different purposes.

Fig. 1: The protractor lies on the ground. A straight rod is affixed with play dough to the center of the semicircle, perpendicularly to the plane of the protractor. The Sun throws a shadow of the rod. If this measurement is done at High Noon (when the Sun is highest in the sky) then the shadow is exactly in the North-South direction. Turn the protractor until the shadow falls through the 0° mark. East is then at the 90° mark. This arrangement can be used to determine the N-S direction, necessary for other measurements. Once you have the N-S direction, you have established the azimuth angles. See Unit 5.1.

An azimuth measurement is done in Fig. 2. North was determined as shown in Fig. 1. The N-S direction was marked on the base. The Azimuth angle is explained in Unit 5.1 and 5.2. If you are not certain at what time of the day the Sun is highest in the sky (High Noon) consult your instructor or read Unit 5.4.

The altitude of an object can be measured with an adaptation of the protractor as shown in Fig. 3. Attach a straw to the protractor. Find out the best length for

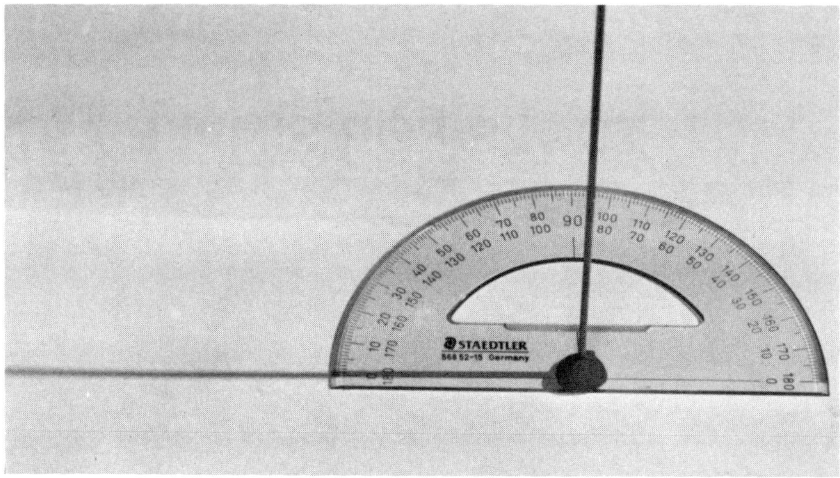

Figure 1

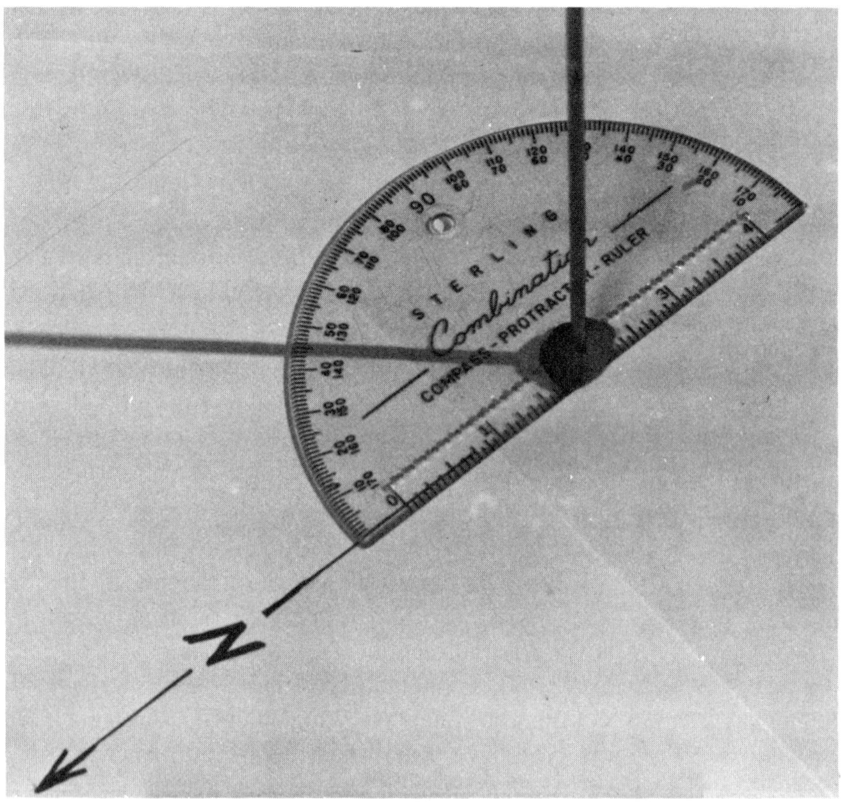

Figure 2

good sighting. **NEVER LOOK THROUGH SUCH A DEVICE AT THE SUN.** If you need to measure the altitude of the Sun, do it by projection, as shown in Unit 5.3, p. 5–13. The line with plumb bob should be able to swing freely.

For some other angle measurements you may wish to attach a sighting device as shown in Fig. 4 and mount the protractor horizontally. You may have the sighting straw mounted so that it can rotate, or the whole assembly may be rotated.

EXERCISES

Do the measurements as specified by your instructor on p. 3–5.

Figure 3

Figure 4

Unit 3.1 Name ______________________________

Date due ____________ Instructor ______________________ Section __________ Seat ____

1) Use your protractor with sighting device to measure the height of

a) ______________________ b) ______________________

Note: If you are at the same level as the base of a building, for example, and the top of the building appears at an angle of 45° with the level, as seen through your sighting device, then the building is as high as the distance from its base to the point of measurement. See Unit 7.4, p. 7–37 for more of an explanation.

The height of a) ______________________ is: ____________

The height of b) ______________________ is: ____________

2) ______________________________

____

____

____

3) ______________________________

____

____

____

4) ______________________________

____

____

____

Unit 3.2
The Cross-Staff

OBJECTIVE To construct a cross-staff for use in making celestial measurements.

INTRODUCTION The cross-staff is a primitive form of the mariner's sextant. As the name implies, it is a simple instrument consisting of a calibrated stick (the staff) and a movable transversal (the cross). The instrument can be used to measure the angle subtended at the eye by distant objects. The advantage of the cross-staff over the better known quadrant or astrolabe is that it does not require a plumb bob to determine the zenith.

The cross member should be made with a specific length. The one shown has two lengths available, 10 cm and 20 cm if you are to use a meter stick, and 4 inches and 8 inches if you are to use a yard stick.

This cross is made so that it can be attached to your staff, either a yard or meter stick, in such a manner that it will slide along the staff. To make measurements, you place the staff with its 0 mark on your cheekbone just below the eye and sight along the staff while adjusting the position of the cross until the two objects line up with the edges of the cross, shown in Fig. 1, and read the distance from your eye to the cross. You then look up the staff reading in the table to determine the angle between the objects.

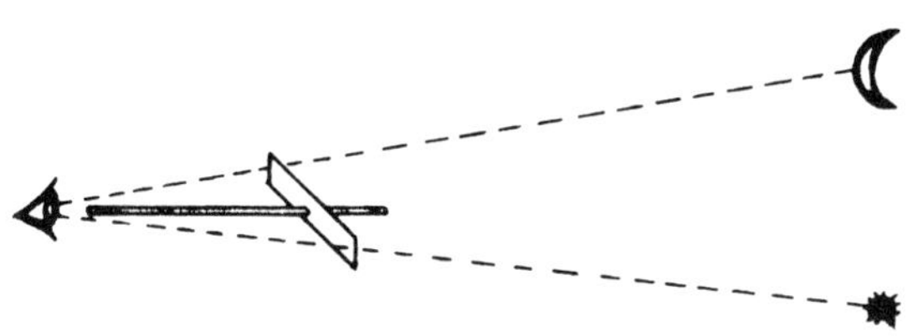

Figure 1

The cross staff pattern shown in Fig. 7 on p. 3–21 may be cut for 10 and 20 cm or for 4 and 8 inches. This pattern should be transferred to poster board and cut to the correct size. The dotted lines indicate where the cross is to be folded around your staff. The two tabs indicated by A are to be folded along the top of your staff to make the cross rigid. A medium weight rubber band is placed around the cross to hold the folds and braces to your staff. The cross should slide up and down the staff, and your readings should be taken at the end of the cross. This reading may then be converted to degrees by using the table at the end of this unit.

When this type of measurement is used with your star maps, a number of rather interesting projects may be done. The star maps are a special projection of the sky. The star maps are plotted on a simple square grid, and hence for reasonably small angles you may use length to represent the angle. On these maps 1 cm = 5° or 2 mm = 1 degree. The stars plotted on these sheets are generally the brightest and easiest to see. By measuring the angle between an object (like the moon or a planet) and three known stars, its position can be determined. If you were plotting the motion of the moon against the background of fixed stars you need, at most, only three angles (sometimes two will do) to determine where to plot the object on the star maps.

To plot the position of a body on the star maps you identify two or three stars on the maps near the object. The angle between the object and each of the stars is then determined by the method described above. The angle from the object to the first star is set on a compass.

The easiest way to do this is to set your compass with the scale at the side or bottom of your star maps. An arc is drawn from the first star with the compass. The same procedure is used with the other two stars. Where all three arcs cross will be the position of the object. Because this is an approximate method you may find that your arcs form a small triangle. Use the center of the triangle as the position of the object.

EXAMPLE

Suppose the object you wish to plot is in the area of Leo, Canis Major, and Gemini, and you have identified the three reference stars Regulus, Procyon, and Pollux. Using the 10 cm cross-staff you read 18 cm on the staff for the measurement from Regulus, 31.6 cm from Procyon, and 63.5 cm from Pollux. These readings correspond to the following angles: Regulus-object 31°, Procyon-object 18°, and Pollux-object 9°. These angles were obtained by simply reading the table. A compass is then used to draw the three arcs shown in Fig. 2. The arc from Regulus is drawn by using the scale on the star map to set the compass for 31°. The other two arcs are drawn in the same manner. Where all three cross will be the location of the object. Note that you could now read the location of the object as R.A. = 122°, Declination = +21°.

EXERCISES

The following will allow you to gain experience with your cross-staff. At least one of the following areas of the sky should be visible any time of the year. Use your cross-staff to determine the angles from the reference stars A,B,C to the object (in this case a star). Record the staff readings and the angles. Draw the arcs from the reference stars to the star in question to find out how close you can come to the true position of the star.

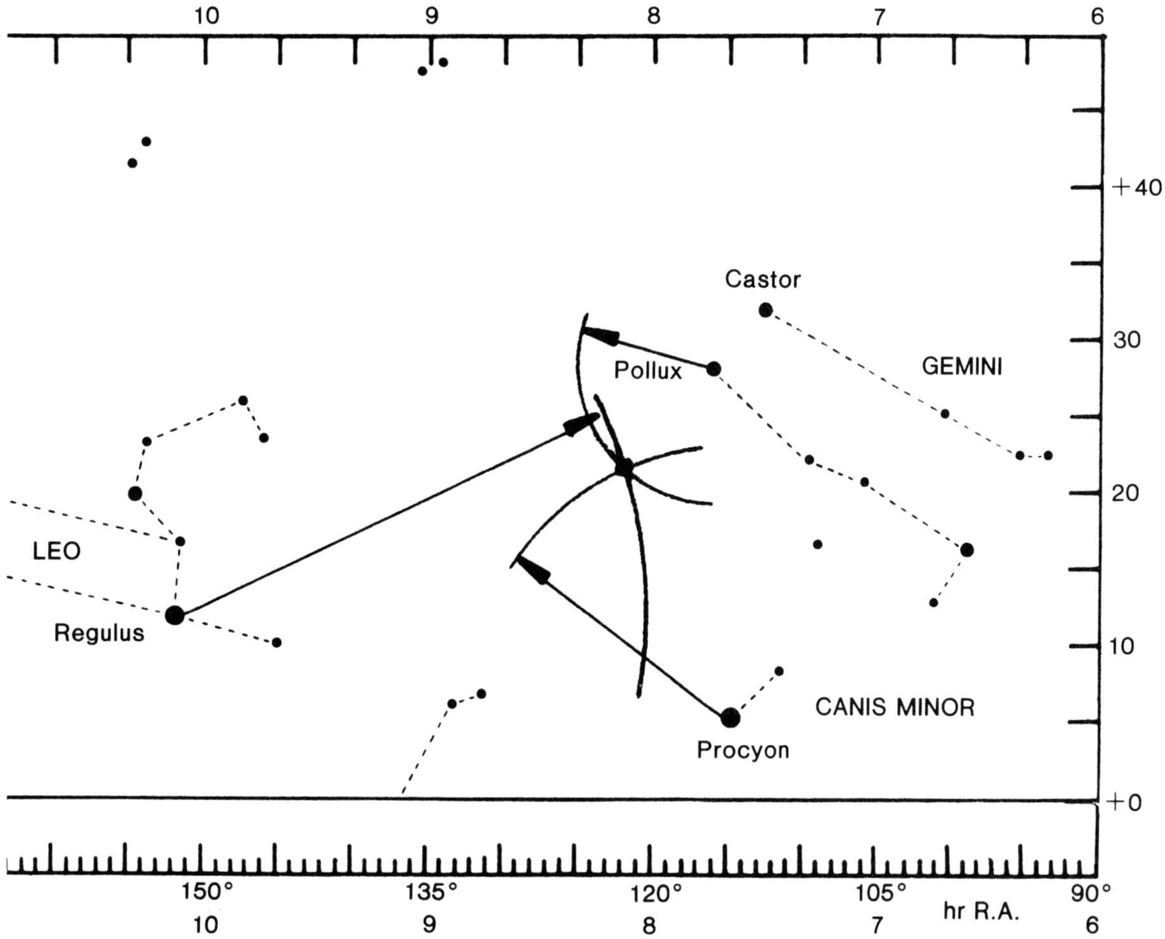
10
9
8
7
6
+40
30
20
10
+0
Castor
Pollux
GEMINI
LEO
Regulus
CANIS MINOR
Procyon
150°
135°
120°
105°
90°
10
9
8
7
6
hr R.A.

Figure 2

Unit 3.2 Name ____________________

Date due __________ Instructor ____________________ Section __________ Seat _____

1) Object: Aldebaran

		Staff reading	Angle
Reference Stars	A. (Betelgeuse)	________	________
	B.	________	________
	C. (Rigel)	________	________

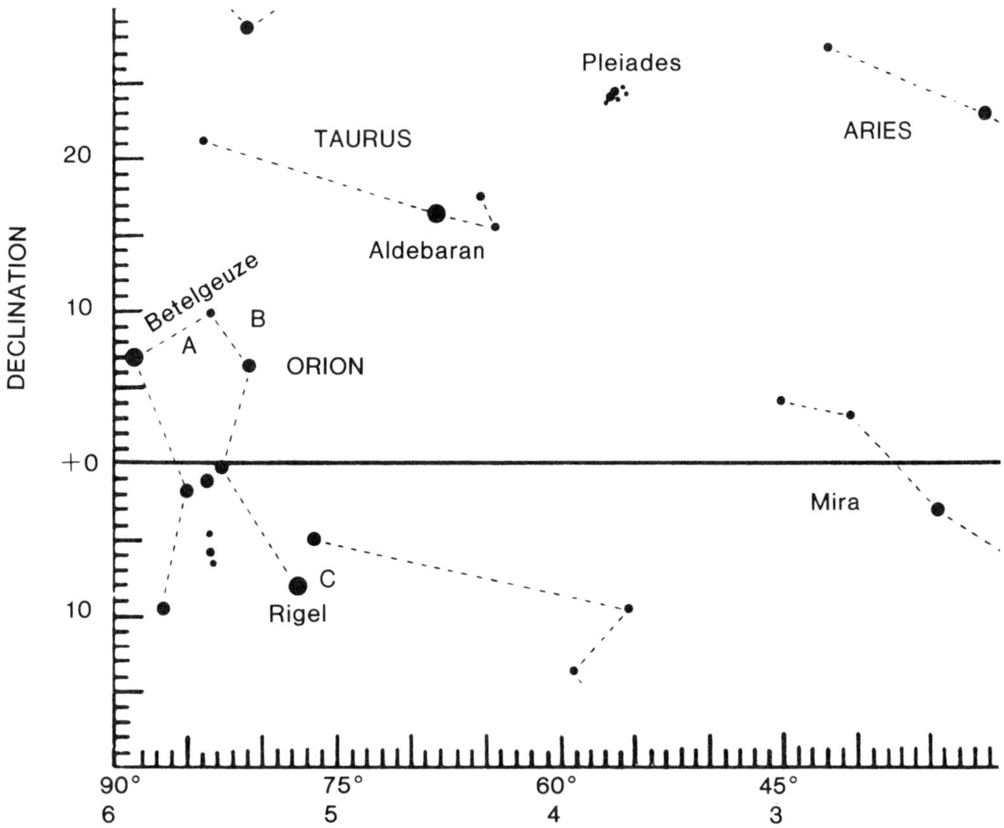

Figure 3

Unit 3.2 Name ______________________________

Date due ____________ Instructor ____________________ Section ________ Seat ____

2) Object Procyon

		Staff reading	Angle
Reference Stars:	A. (Regulus)	____________	____________
	B. (Pollux)	____________	____________
	C. (Sirius)	____________	____________

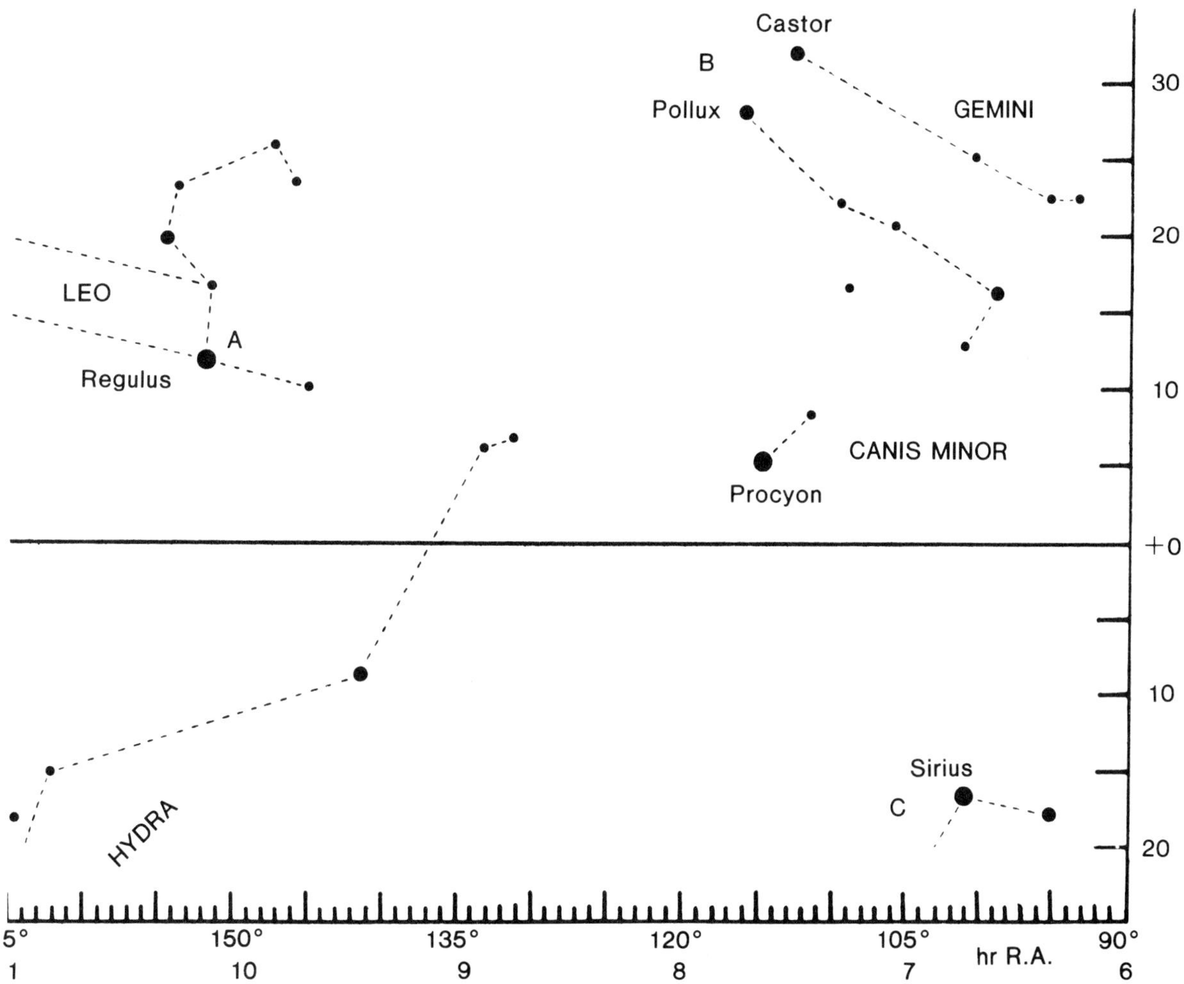

Figure 4

Unit 3.2 Name ______________________________

Date due ____________ Instructor ____________________ Section ________ Seat ____

3) Object Antares

		Staff reading	Angle
Reference stars:	A.	________	________
	B.	________	________
	C.	________	________

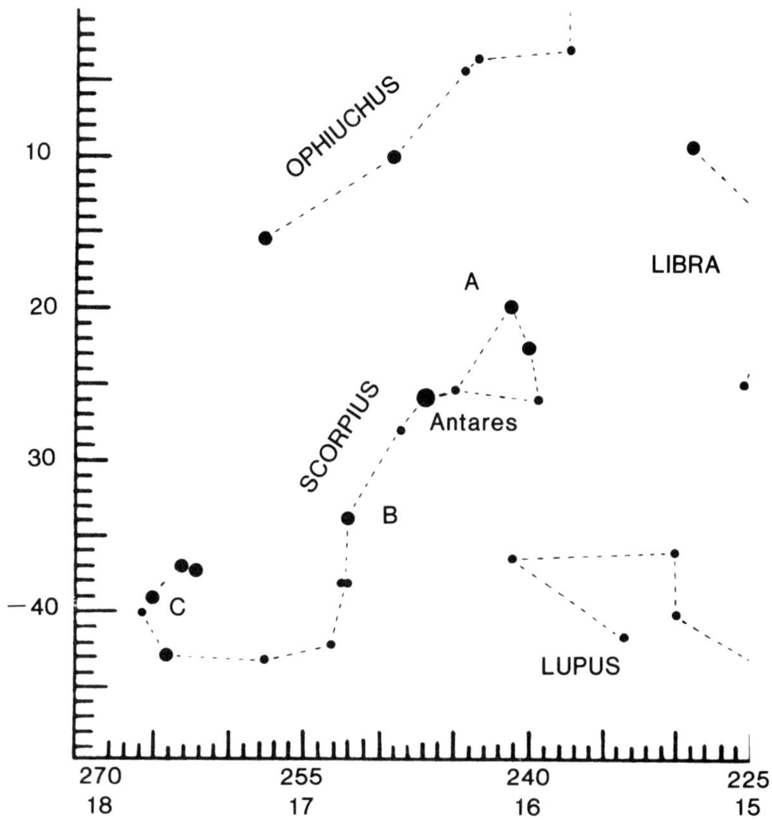

Figure 5

Unit 3.2 Name ______________________________

Date due ____________ Instructor ______________________ Section ________ Seat ____

4) Object: Center star of the cross in Cygnus

		Staff reading	Angle
Reference stars:	A.	______	______
	B. (Deneb)	______	______
	C. (Vega)	______	______

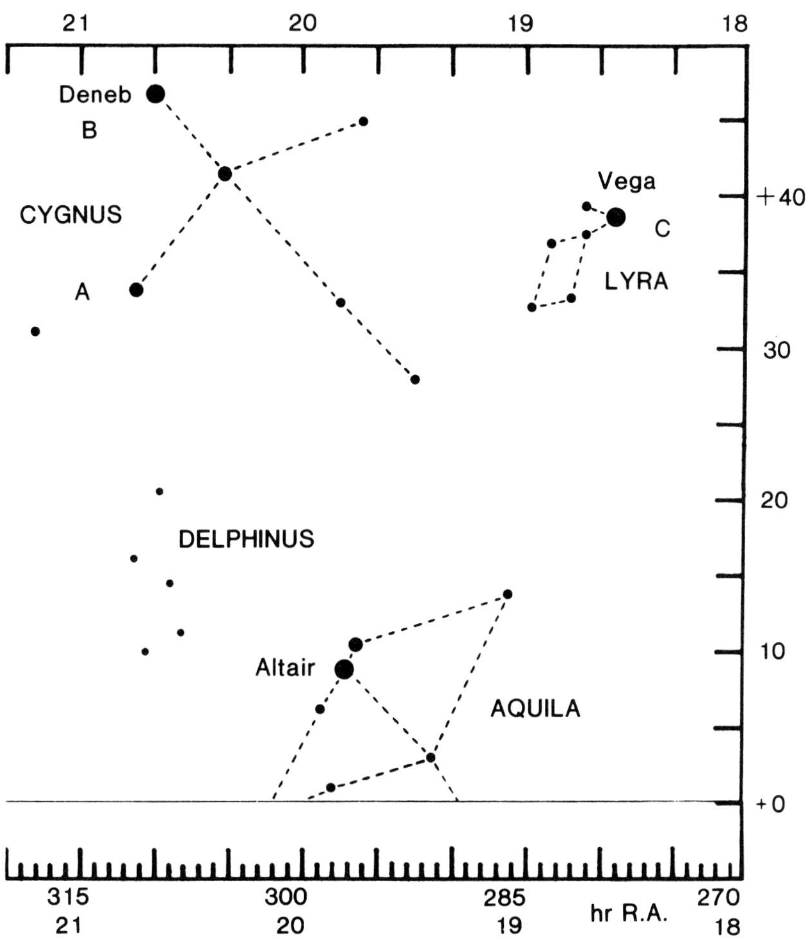

Figure 6

Angle in degrees	STAFF READING in cm for 10 cm cross	in inch for 4 in cross	in cm for 20 cm cross	in inch for 8 inch cross
6.0	95.4	37½		
6.5	88.0	34⅝		
7.0	81.8	32¼		
7.5	76.3	30		
8.0	71.5	28⅛		
8.5	68.1	26¾		
9.0	63.5	25		
9.5	60.2	23¾		
10.0	57.1	22½		
10.5	54.4	21⅜		
11.0	51.9	20½		
11.5	49.7	19½	99.3	39
12.0	47.6	18¾	95.1	37½
12.5	45.7	18	91.3	36
13.0	43.9	17¼	87.8	34½
13.5	42.2	16⅝	84.5	33¼
14.0	40.7	16	81.4	32
14.5	39.3	15½	78.6	31
15.0	37.9	14⅞	76.0	29¾
15.5	36.7	14½	73.5	29
16.0	35.6	14	71.1	28
16.5	34.5	13⅝	69.0	27¼
17.0	33.6	13¼	67.1	26½
17.5	32.5	12¾	65.0	25½
18.0	31.6	12½	63.1	25
18.5	30.7	12⅛	61.4	24¼
19.0	29.9	11¾	59.6	23½
19.5	29.1	11½	58.2	23
20.0	28.4	11⅛	56.7	22¼
20.5	27.6	10⅞	55.3	21¾
21.0	27.0	10⅝	54.0	21¼
21.5	26.3	10⅜	52.7	20¾
22.0	25.7	10⅛	51.4	20¼
22.5	25	9⅞	50	19¾
23.0	24.6	9⅝	49.2	19¼
23.5	24.9	9¾	48.1	19½
24.0	23.5	9¼	47.0	18½
24.5	23.0	9	46.1	18
25.0	22.5	8⅞	45.1	17¾
25.5	22.1	8¾	44.2	17½
26.0	21.7	8½	43.3	17
26.5	21.2	8⅜	42.5	16¾
27.0	20.8	8¼	41.6	16½
27.5	20.4	8	40.9	16
28.0	20.0	7⅞	40.1	15¾
28.5	19.7	7¾	39.4	15½
29.0	19.3	7½	38.7	15
29.5	19.0	7½	38.0	15
30.0	18.7	7⅜	37.3	14¾
30.5	18.3	7¼	36.7	14½

STAFF READING (Continued)

Angle in degrees	in cm for 10 cm cross	in inch for 4 in cross	in cm for 20 cm cross	in inch for 8 inch cross
31.0	18.0	7⅛	36.1	14¼
31.5	17.7	7	35.5	14
32.0	17.5	6⅞	34.9	13¾
32.5	17.2	6¾	34.3	13½
33.0	16.9	6⅝	33.8	13¼
33.5	16.6	6½	33.2	13
34.0	16.4	6½	32.7	13
34.5	16.1	6⅜	32.2	12¾
35.0	15.9	6¼	31.7	12½
35.5	15.6	6⅛	31.2	12¼
36.0	15.4	6	30.8	12
36.5	15.2	6	30.3	12
37.0	14.9	5⅞	29.9	11¾
37.5	14.7	5¾	29.5	11½
38.0	14.5	5¾	29.0	11½
38.5	14.3	5⅝	28.6	11¼
39.0	14.1	5½	28.2	11
39.5	13.9	5½	27.8	11
40.0	13.7	5⅜	27.5	10¾
40.5	13.6	5⅜	27.1	10¾
41.0	13.4	5¼	26.7	10½
41.5	13.2	5¼	26.4	10½
42.0	13.0	5⅛	26.0	10¼
42.5	12.9	5⅛	25.7	10¼
43.0	12.7	5	25.4	10
43.5	12.5	5	25	10
44.0	12.4	4⅞	24.8	9¾
44.5	12.2	4¾	24.4	9½
45.0	12.1	4¾	24.2	9½

Note: Remove the dark areas for the 10, 20 cm cross.

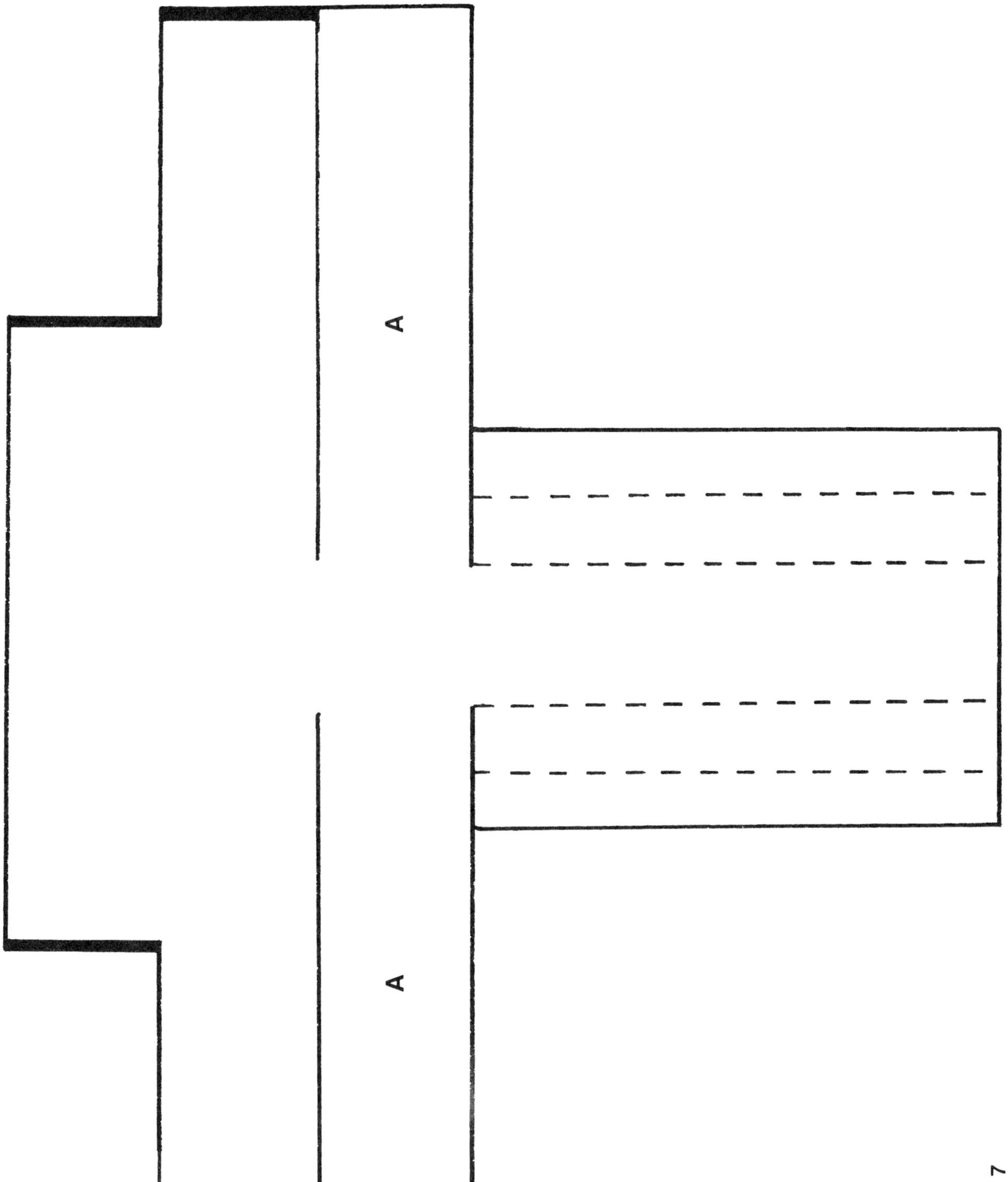

Figure 7

Unit 3.3
Use of the Starfinder

OBJECTIVE To use the starfinder for the measurement of a) Declination and Right Ascension (R.A. and Dec) b) Altitude and Azimuth.

APPARATUS A commercial starfinder or the self-made starfinder of Unit 3.4.

THEORY The RA and Dec coordinates are explained in Unit 5.5. Altitude and Azimuth coordinates are explained in Unit 5.1. A brief repetition: The Dec of a celestial object is its angle from the celestial equator measured along its hour-circle. The Dec is (+) if the object is located in the Northern celestial hemi-sphere or above the celestial equator; if "below" the celestial equator the Dec is negative (−). The RA of a celestial object is measured eastward along the celestial equator from the vernal equinox, and ranges from 0° or 0 hr to 360° or 24 hr.

The altitude-azimuth coordinate system uses the observer's horizon as the reference plane. The altitude of an object is the angle above the horizon, ranging from 0° for the horizon to 90° for the zenith. The azimuth is the angle measured eastward along the horizon ranging from 0° for the North point to 360° all the way around back to the North point. The compass points are then: East 90°, South 180°, West 270°.

PROCEDURE

A) Measuring Declination and Right Ascension

A commercial starfinder, known as the RoVi* Starfinder is shown in Fig. 1. The starfinder consists basically of two scales that can be aligned with coordinates on the celestial sphere. The semi-circular scale is your Dec scale and the circular scale is used for the RA.

1) To set up the starfinder, you use a level base, e.g. you may use a camera tripod to hold a surface to be leveled or any other device you can think of, for example, a board with legs on one end, with the other end in the window of your car. By raising or lowering the window you can level the board.
2) Set the viewing tube to read 90° on the Dec scale as shown in Fig. 2.
3) Loosen the wing nut so that the complete assembly can rotate around its axis. With the fingers of one hand under the RA scale (the circular one) rotate the assembly around the axis and rotate the complete starfinder about a vertical axis until you have Polaris centered in the viewing tube.
4) When Polaris is centered in the viewing tube, lock your tripod or tape the base of the starfinder to your level working surface and tighten the wingnut. Check the following four items:
   a) is Polaris centered in the tube?
   b) does the Dec scale still read 90°?
   c) have you noted the end of the viewing tube that you are using? (It is good practice to use the end flush with the arm)
   d) is the base of the starfinder securely fixed to the tripod or to your working surface?

*RoVi Plastics, Inc., 2317 Blue Smoke Ct. N., Ft. Worth, TX 76015.

Figure 1

Figure 2

5) At this point you have the RA scale parallel to the celestial equator and you have defined the polar axis. The Dec scale will now read correctly the Dec of any celestial object if centered in the viewing tube. You should note the Dec scale has 0° at the center and 90° at both ends. To determine the sign (+ or −) of the Dec you need only to check the line of sight relative to the RA scale as shown in Fig. 3. If the line of sight is above the edge of the RA scale you are looking North of the equator and the sign will be positive (+). If your line of sight falls below the edge of the RA scale you are looking South of the equator and the Dec will be negative (−).
6) To set the RA scale it is necessary to use a reference star of known coordinates. The following table gives a list of reference stars used to set your RA scale.

Reference Star	**RA**	**DEC**
Aldebaran in Taurus	68°	+16°
Betelgeuze in Orion	88°	+ 7°
Sirius in Canis Major	101°	−17°
Regulus in Leo	151°	+12°
Spica in Virgo	201°	−11°
Antares in Scorpius	247°	−26°
Altair in Aquila	297°	+ 9°

With one of the reference stars centered in the viewing tube, rotate the RA scale until the correct RA appears under the indicating tab on the Dec scale. Fig. 4 shows a RA reading of 210°.
7) Also check the Dec reading for your reference star. If this reading is not very near the Dec values listed in the table you have not set up the starfinder correctly.
8) When you have completed the above procedure and center another celestial object in the viewing tube by carefully moving the viewing tube arm and rotating the Dec scale, your starfinder will read the coordinates of that celestial object.
9) Because the earth rotates on its axis at a rate of 1° every 4 minutes, it is necessary to keep track of how much time has elapsed once you set the RA scale with the reference star. For the instrument to give correct readings the RA scale must be rotated regularly. You may do this by increasing the RA scale reading under the indicating tab 1° for every 4 minutes (actually it is generally easier to work 8–10 minutes and increase the reading 2°) or you may center on a reference star just before each measurement. If your measurement is done within 10 minutes after centering on a reference star your measurement is generally acceptable.

B) Measuring Altitude-Azimuth

To set up the starfinder for these measurements:

1) You need a level working surface (either a board leveled on a tripod or some other level surface). You must align the circular RA scale with the horizon. This can be done by using a small level or if your working surface is quite level, by making sure that all points of the bottom of the RA scale are 3⅛ inches above the base. For the self-made starfinder it

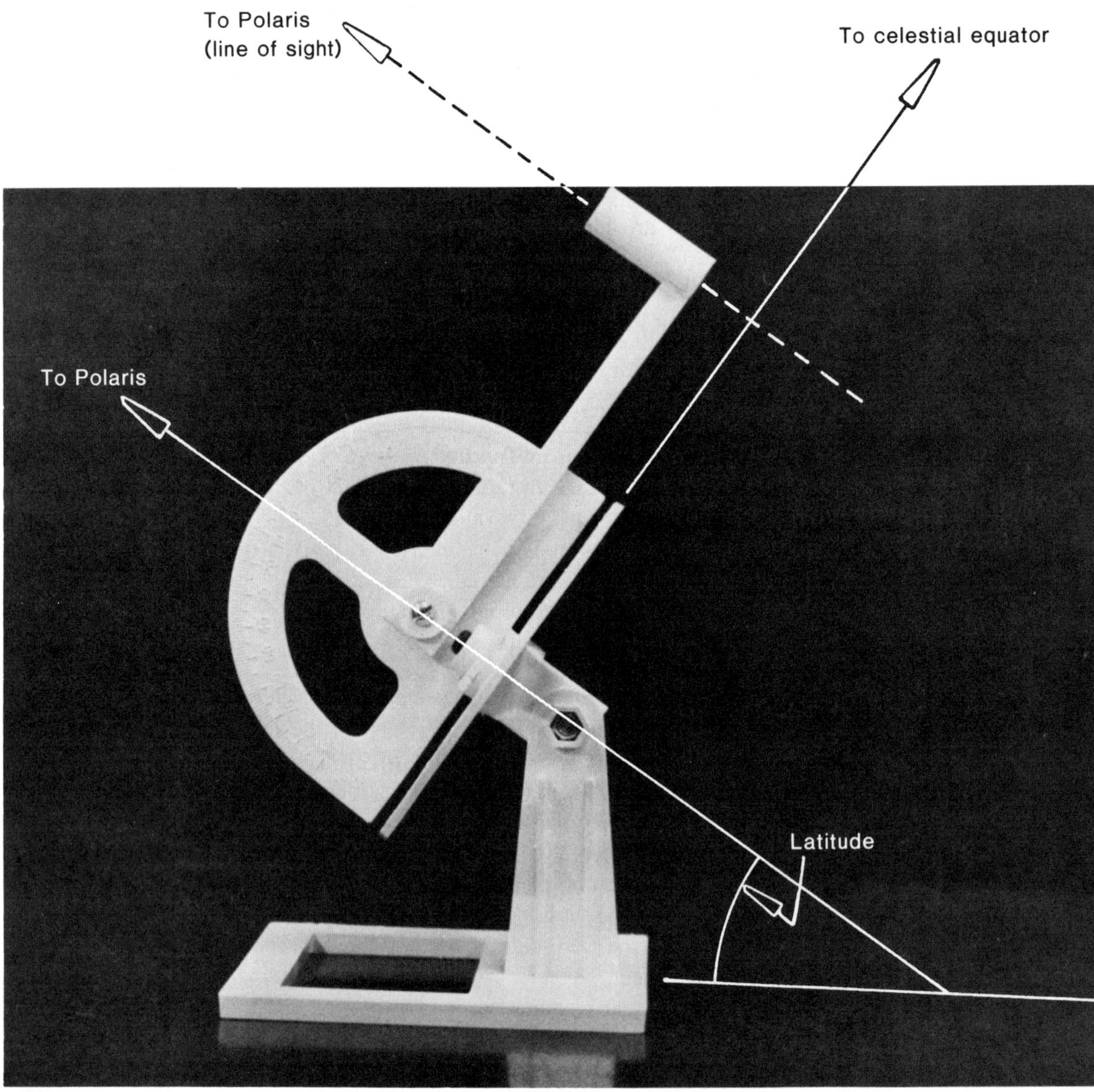
To Polaris
(line of sight)
To celestial equator
To Polaris
Latitude

Figure 3

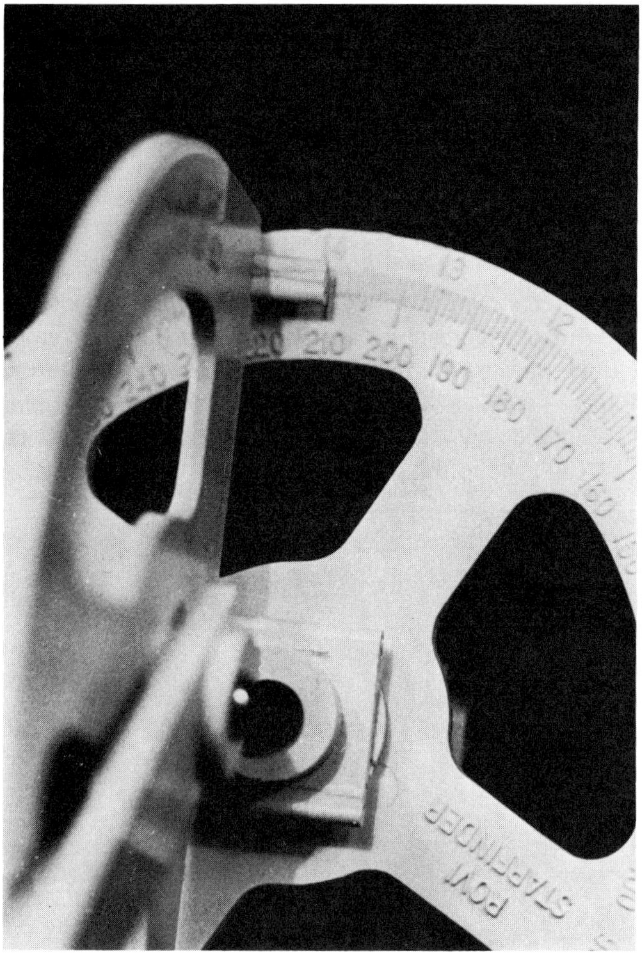

Figure 4

should be constructed so that the circular scale is parallel to the base rather than parallel to the equator.

2) When you have the circular scale level, to determine the North point with azimuth 0°, while working at night, center Polaris in the viewing tube. For day work you must know the true North direction either by finding North at night with Polaris or by considering the position of the Sun. The Sun is exactly South at local noon, which is 12:00 on your clock if you live on the 75°, 90°, 105°, or 120° meridans. Consult your instructor or see Unit 5.4.

   Once you have the circular scale level and the viewing tube pointing true North or South, with the RA indicating-tab North, rotate the RA scale until 0° is under the indicating tab.

3) To determine the altitude and azimuth of any celestial object, simply center the object in the viewing tube (be careful not to change ends of the viewing tube). The reading on the Dec scale will be the altitude of the object. The reading on the RA scale does **NOT** give you the correct azimuth value. Both the RA and azimuth scale read from 0° to 360°, they do so in opposite directions. Hence your azimuth value will be: azimuth = 360° − RA scale reading.

4) If you are measuring very bright objects like the Moon or Sun, you can easily center these objects in the viewing tube by projecting the shadow of the viewing tube on a piece of paper. One **MUST** do this when measurements of the Sun are being made. You must never look directly at the Sun because it may permanently damage your eyes.

Exercises using the starfinder for RA and Dec measurements are Units 5.6, 5.7, 5.8 and 5.9. Exercises using the starfinder for altitude-azimuth measurements are Units 5.1 and 5.3.

An interesting measurement not covered in any of the above units can be made at night. If you set the starfinder to read altitude-azimuth, note the reading on the Dec scale when you have Polaris centered in the viewing tube, and record this number. ______________ °. If the starfinder is set up correctly this should be your geographic latitude.

Unit 3.4

Making Your Own Starfinder

OBJECTIVE To construct a starfinder suitable for R.A. and Dec. or azimuth and altitude measurements.

INTRODUCTION If the commercial starfinder is not available, you may construct a usable starfinder from cardboard cut from an ordinary box. However, if three-quarter inch styrofoam is available the starfinder may be constructed with this material somewhat quicker in that one layer of styrofoam may be used for two layers of cardboard.

APPARATUS Knife, scissors, glue, soda straw, nail (1.5 inch), and cardboard.

PROCEDURE

1) Cut the pattern for the support of your starfinder for your latitude by using the template of Fig. 2. Note: If you are using cardboard you will need two of these.
2) Take the nail and glue the two parts of this support together with the sharp end of the nail protruding approximately one inch above the slanted surface as shown in Fig. 3. You may wish to make two starfinders. The first template in Fig. 3 is for a starfinder that measures Right Ascension and Declination. The second template is for a starfinder that will measure altitude-azimuth. (See Unit 3.3) Be sure that the nail is as nearly perpendicular to the slanted surface as possible.
3) The base can be a rectangle of double thickness of cardboard. A reasonable size is 3½ inches by 3½ inches.
4) Cut the circular Right Ascension scale of Fig. 4a to size. Cut a single thickness of cardboard to this size and glue the scale to the cardboard.
5) Cut out the rectangular Declination scale of Fig. 3b. Cut two pieces of cardboard the same size as this scale. Glue the cardboard together and then glue the scale to the cardboard.
6) Cut the pattern for the viewing tube from Fig. 3c. Use this to cut the viewing arm from a single thickness of cardboard. Glue about 2 inches of the soda straw to the end of this support in the position indicated by the dashed lines in Fig. 4c.

Once all parts are cut out and the scales are glued to the proper parts you should note from the photograph in Fig. 1 that the arm for the viewing tube is attached to the declination scale with a small screw. Generally a number 2–56 is best, however, almost any small screw that will allow the viewing tube arm to move yet retain its final position will do. Also note that the nail is pushed through the center of the Right Ascension scale and into the Declination scale just below the small screw. Care should be taken to insure that both scales can be rotated individually without moving the other scale.

The use of this starfinder is the same as the commercial starfinder of Unit 3.3.

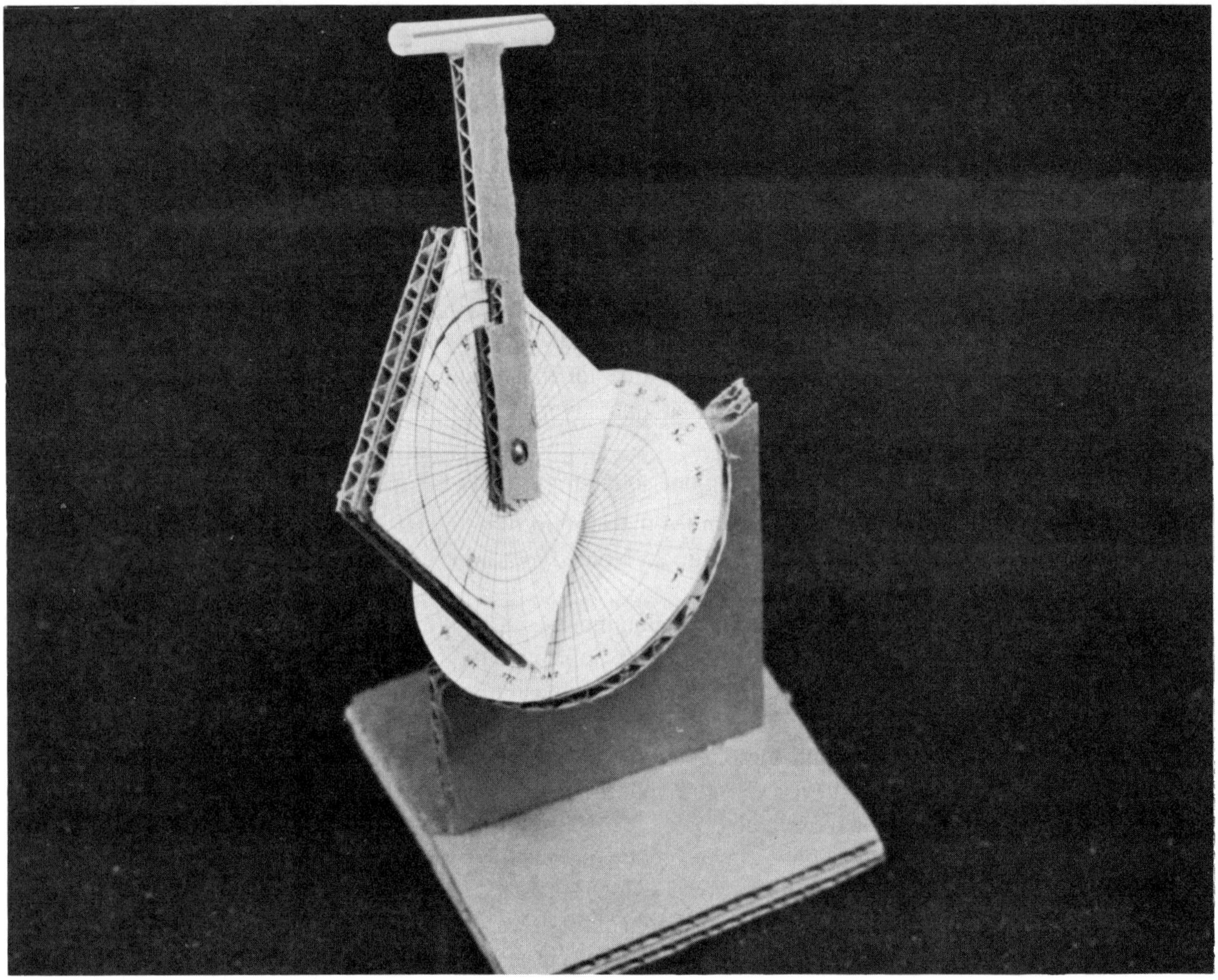

Figure 1

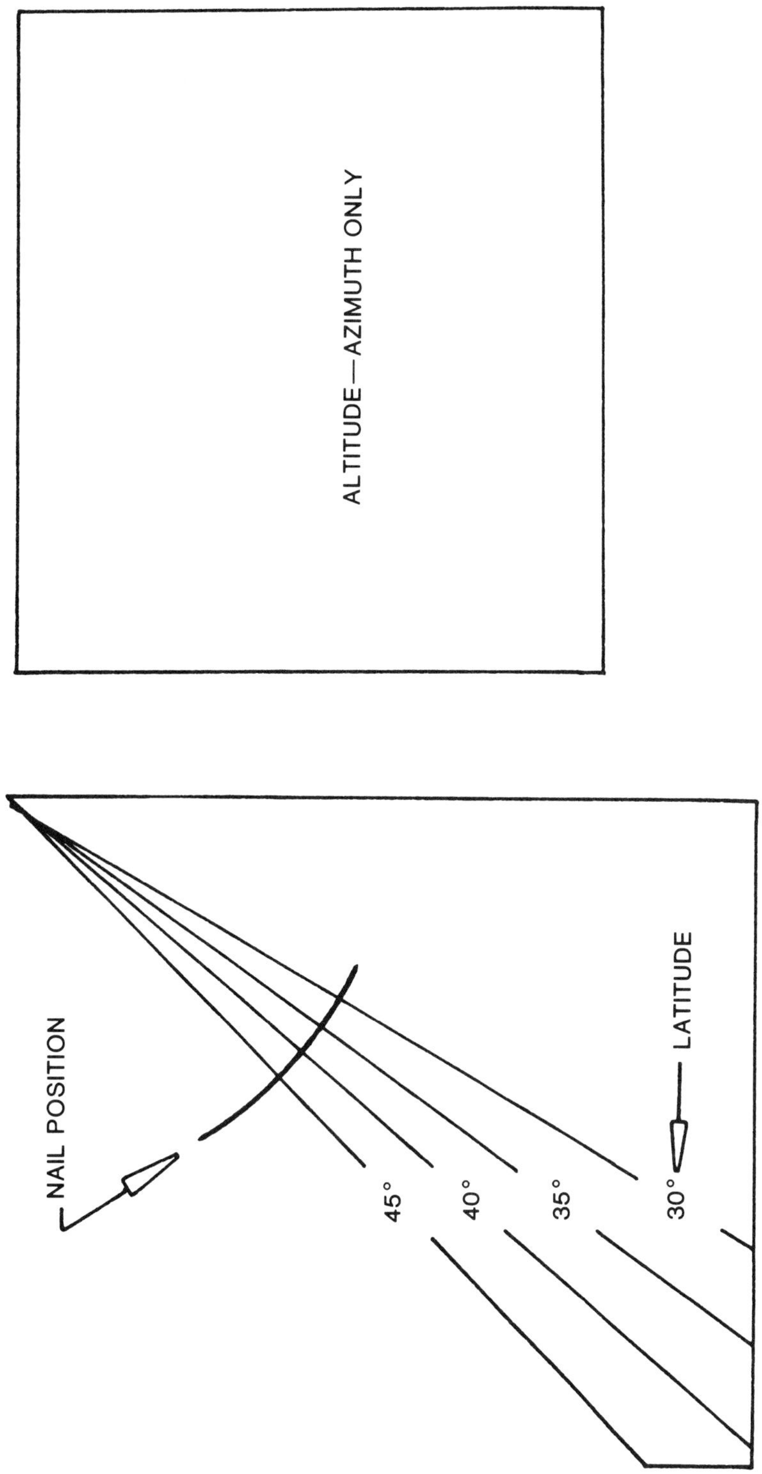

Figure 2

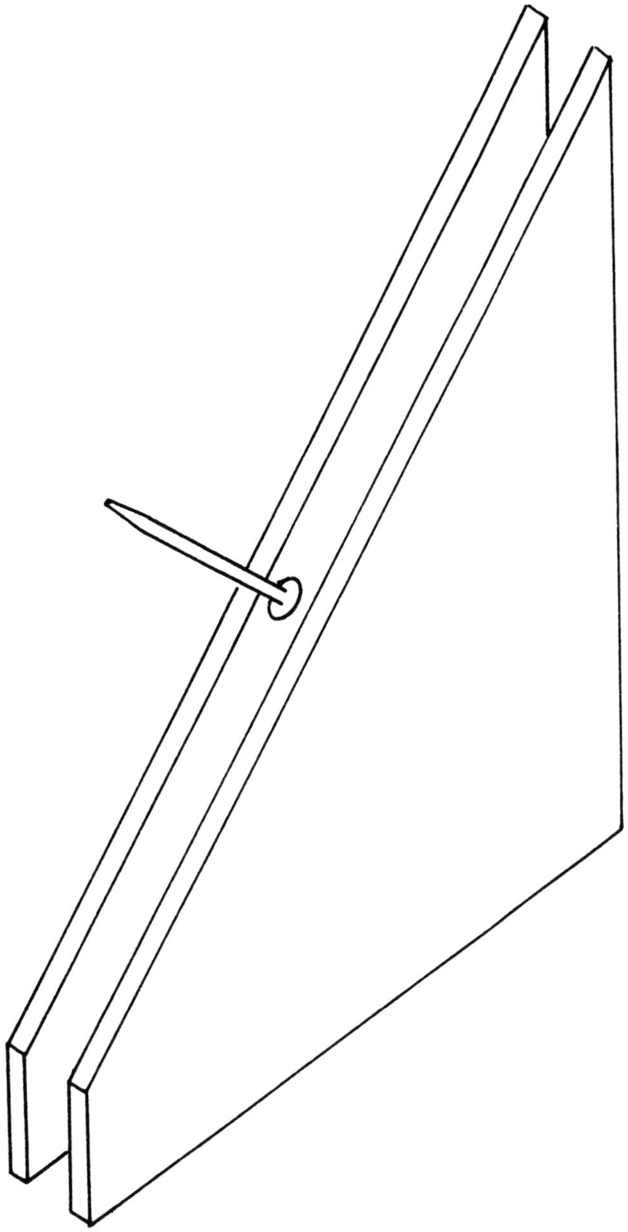

Figure 3

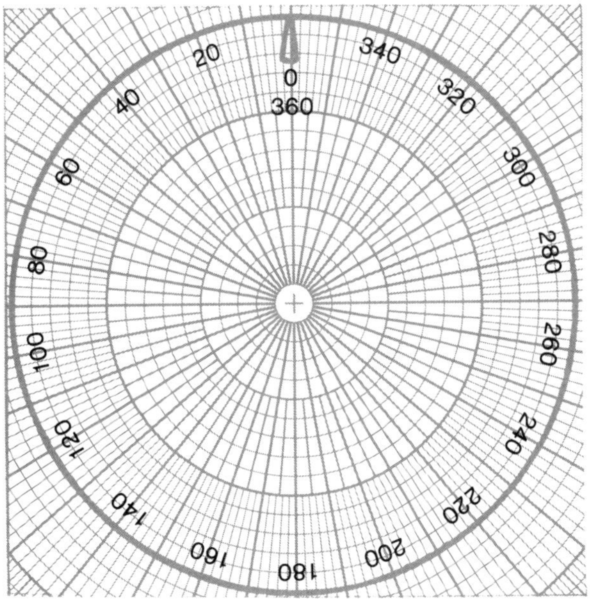

Figure 4a. R.A. Scale

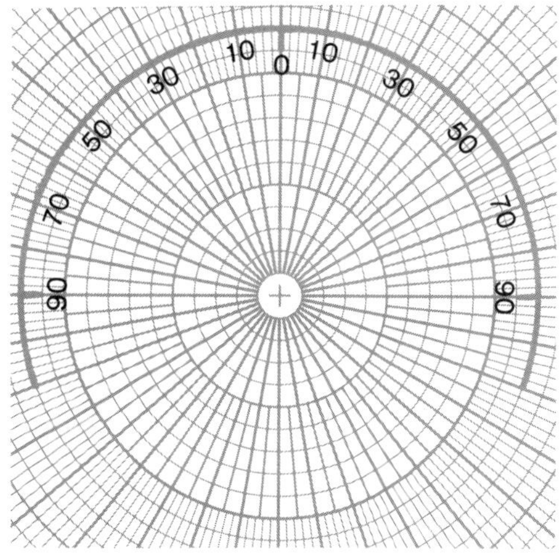

Figure 4b. Dec Scale

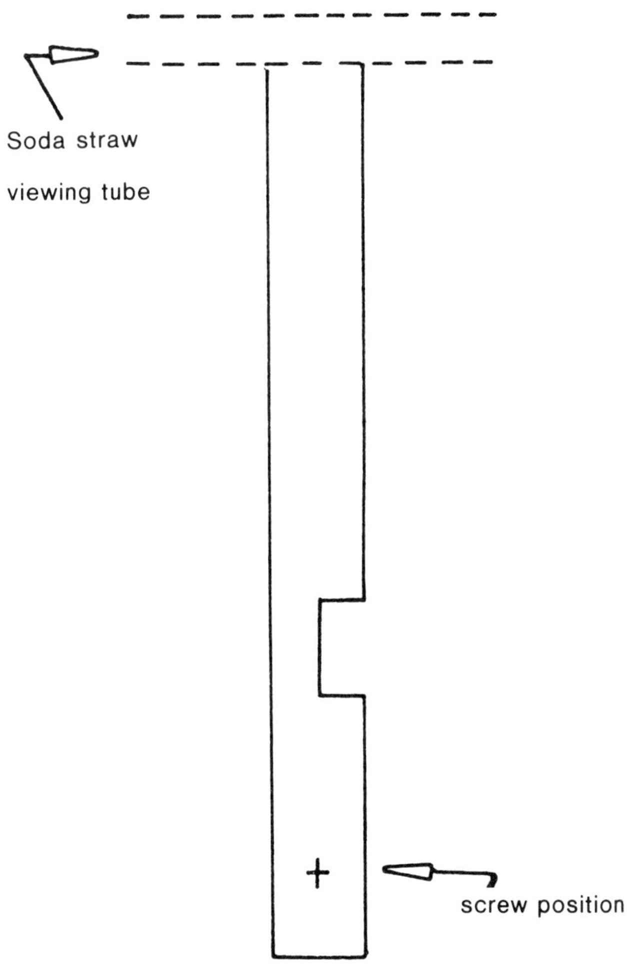

Figure 4c. Viewing Tube Support

Unit 3.5

Astronomical Telescopes

OBJECTIVE

To become familiar with magnification, light-gathering power, and resolving power of telescopes.

INTRODUCTION

The basic astronomical telescopes are of two types: Refracting and reflecting. In both types light is gathered by one element (called the objective) and the resulting image is magnified by a second element, a small lens, (called the eyepiece).

You should note that the light gathering element of the refractor is a lens, while the light gathering element for the reflector is a mirror. The basic design of the refractor type telescope is shown in Fig. 1. The basic design of the reflector type is shown in Fig. 2.

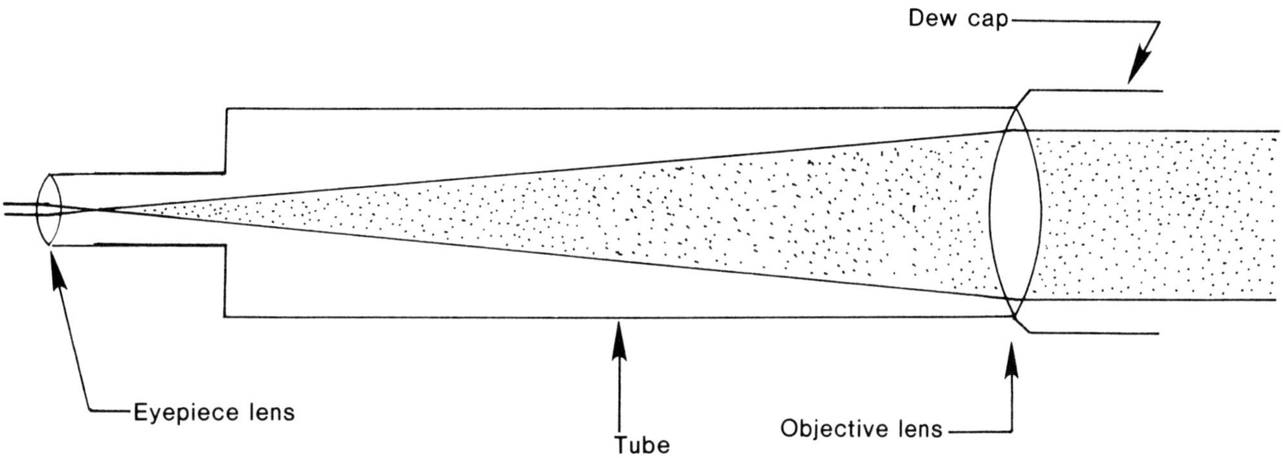

Figure 1. The Refracting Telescope

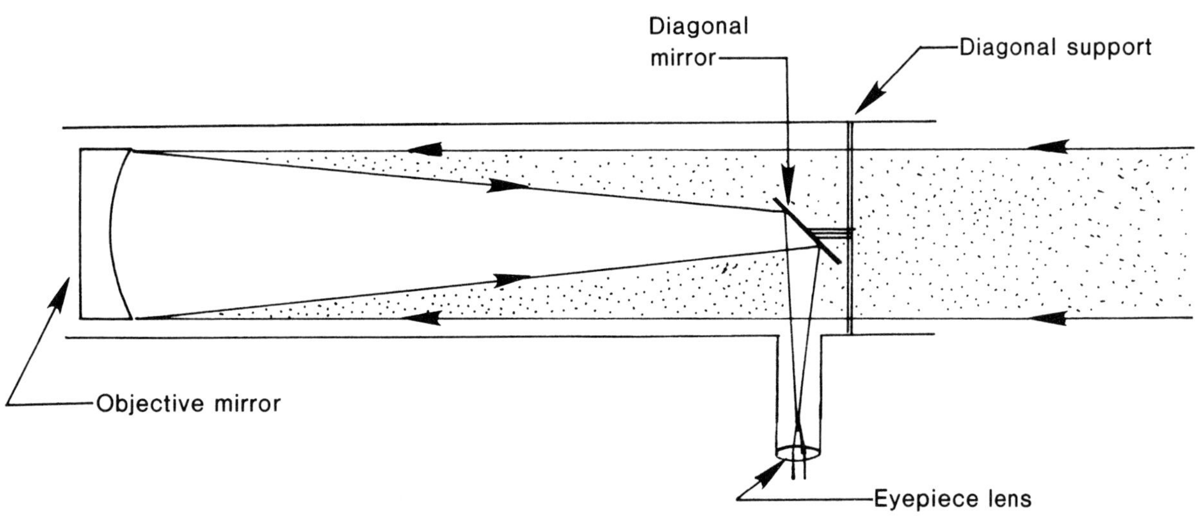

Figure 2. The Reflecting Telescope

When an astronomical telescope forms an image the quality of the image is referred to in terms of size, brightness, and resolution. There are a number of simple rules that will allow you to estimate the quality of the image formed by the optical instrument. Any objective element, lens or mirror, will have a focal length. This is simply the distance from the element to the point where parallel light would be focused by the element. The size or magnification of the image may be calculated by dividing the focal length of the objective by the focal length of the eyepiece.

$$M = \frac{\text{F.L. (objective)}}{\text{F.L. (eyepiece)}}$$

Note this number has no units. It is generally written as a number followed by an x, i.e. 50x.

One of the first conclusions that most students draw from the above equation is that magnification may be increased without limit. After all you only need to get eyepieces of shorter and shorter focal length to make the magnification increase without limit. The case of unlimited magnification is not feasible. One must refer to magnification as ''usable magnification''. The classical ''rule of thumb'' for the maximum magnification is to multiply the diameter of the objective element by 50. Generally, this ''rule of thumb'' will give you a maximum magnification that is really too large. Using the rule of 50, for an 8 inch telescope the maximum usable magnification would be $50 \times 8 = 400x$, for a three inch telescope this maximum would be 150x.

The brightness of an image, for a given eyepiece, will be determined by the diameter of the objective element. The amount of light gathered by the objective element is determined by the area of that element. The area of a circular element is πR^2 or $\pi D^2/4$ where R is the radius and D is the diameter. Thus, to compare the light gathering power of two instruments you compare the square of the diameter of their objectives. Take as an example a 3 inch telescope and a 6 inch telescope. The squares are $(3)^2 = 9$ and $(6)^2 = 36$. The light gathering power of the 6 inch telescope is four times that of the 3 inch telescope ($36/9 = 4$).

The resolution of an astronomical telescope may be estimated by the following rule. The theoretical ''resolving power'' of a telescope is the angle between two close objects which can just be resolved. This angle (in seconds of arc) is sometimes called the ''Dawes limit''. To estimate this angle take the diameter of the objective element in inches and divide it into the number 6.0. As an example consider the 3 inch and 6 inch telescopes. The 3 inch telescope should (in the theoretical limit) resolve two objects that are $6.0/3 = 2.0$ seconds of arc apart, while the 6 inch telescope should resolve objects that are $6.0/6 = 1.00$ seconds of arc apart.

Unit 3.5 Name ____________________

Date due ____________ Instructor ____________________ Section ____________ Seat ______

Exercises

What two quantities determine the magnification?

The 1) ____________ of the 2) ____________

and the 3) ____________ of the 4) ____________ .

5) Compare the light gathering power of a 2 inch telescope and a 4 inch telescope.

6) Compare the light gathering power of a 2 inch telescope and a 8 inch telescope.

Calculate the resolving power for the following optical instruments.

7) 2 inch ____________

8) 3 inch ____________

9) 6 inch ____________

10) 8 inch ____________

11) Eye (diameter = 0.1 inch) ____________

Unit 3.6
Spectroscopes

OBJECTIVE

To construct a spectroscope suitable for observing spectra from laboratory type sources, and to use the spectroscope to measure a few simple spectra.

INTRODUCTION

The light from most common light sources would be described as white. White light is made up of all colors of light: violet, blue, green, yellow, orange, and red. If you consider red light and blue light, the only difference is a parameter called wavelength. Refer to Fig. 1 and consider this wave as light. You will notice that the part of the wave AB is identical to the part BC. This distance is called the wavelength of the wave. Fig. 2 gives the approximate wavelength ranges for the different colors.

To discuss the wavelengths of visible light a new unit of length is used. This unit is called the angstrom (Å). One angstrom is 10^{-10} meters. Visible light is generally considered as having wave lengths between 4,000Å (violet) and 7,000Å (deep red).

When light from a given source is separated according to wavelength, the resulting separated light is called the spectrum of the source. If the source of light is made from high density material that has been heated, e.g. the common light bulb, the spectrum will contain all colors. This type of spectrum is called ''continuous spectrum'' because all wave-lengths are present. If the source of light is a low density atomic gas the spectrum will not contain all wavelengths but only relative few wavelengths that are characteristic of the atom producing the light. The wave-lengths produced by a given element are so unique that this type of spectrum is sometimes called the fingerprint of the atom.

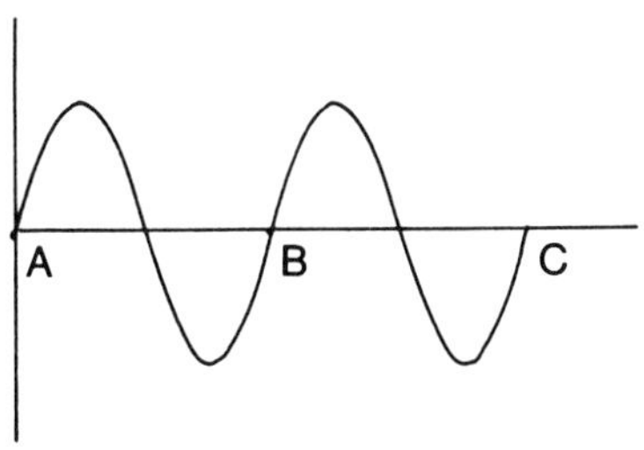

Figure 1

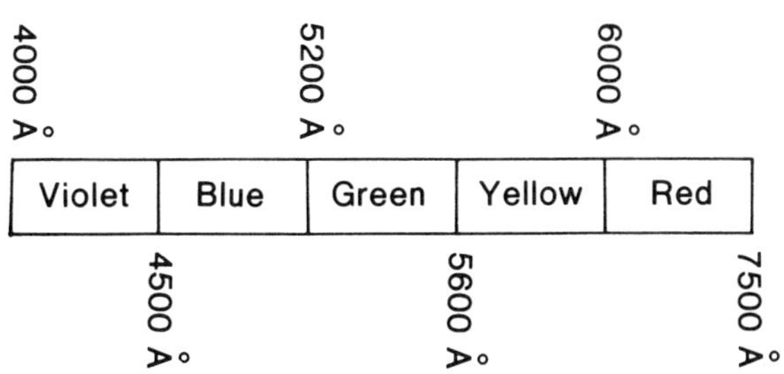

Figure 2

The diffraction grating is a device that uses the phenomenon of diffraction to produce a spectrum. When white light is observed through a grating a spectrum like Fig. 2 will be observed.

APPARATUS

Scissors, razor blade, cement, transmission grating replica* and a shoe box or poster board.

CONSTRUCTING YOUR SPECTROSCOPE

With availability of inexpensive replica grating (less than a dollar) a usable spectroscope can be produced easily. Study Fig. 3. The two patterns for the inside of the spectroscope are shown on pages 3–47 and 3–49. These patterns are designed for use with a shoe box, however they may be glued to poster board and you can make your own box. The pattern also assumes that your grating has 13,400 lines/inch with the length L equal to 25 cm. The necessary modifications for other gratings are given in the step-by-step instructions.

1) Note the number of lines per inch of your diffraction grating. Use the graph in Fig. 4 to determine the grating to slit distance. This distance is the L in Fig. 3.
2) The instrument will be easier to handle if you cut the shoe box so that its depth is two inches.
3) If the length of the shoe box is too long, cut it in half so that one half can be slid into the other half to get the correct length L between the scale and the grating.
4) Cut the patterns as shown on pages 3–47 and 3–49.
5) Glue these patterns to the inside of the shoe box.
6) When your glue has dried, cut the slit with a razor blade. You must be careful with the size of the slit. A slit width of 1/16 inch is generally sufficient. If the slit is too narrow, you will not get enough light. If the slit is too wide the spectra will be fuzzy and it will be difficult to get a wavelength reading.
7) You may use a paper punch to open the viewing hole opposite the slit, or you may cut a quarter inch square hole with the razor blade. This is just to allow you to see inside the spectrometer.

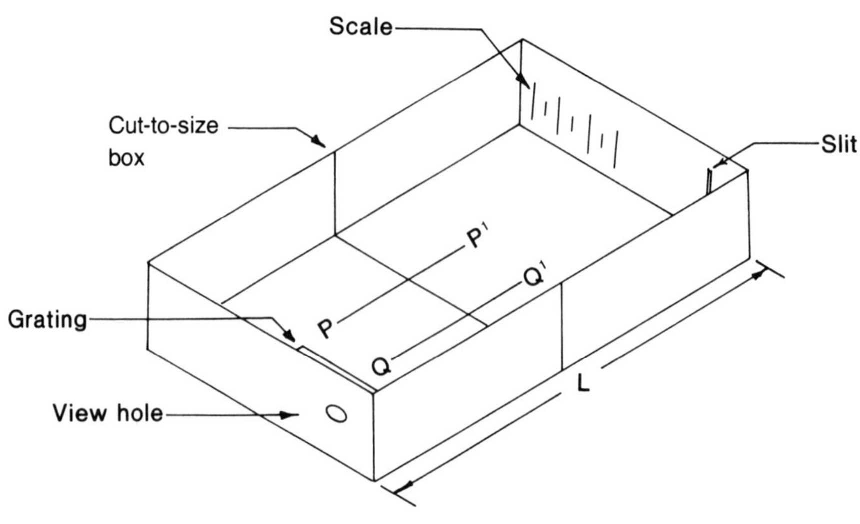

Figure 3

*Edmund Scientific Co., Edscorp Building, Barrington, N.J. 08007.

8) Put the two halves of the box together such that the lines P and Q line up with P′ and Q′. When the length L is correct, tape the two halves together.
9) Tape the grating to the inside of the box over the viewing hole that you cut in Step 7.

You will need a little practice to learn to use the spectrometer. A few good practices can be tried with a regular light bulb. Aim slit end of the spectrometer toward the light and look through the viewing hole. Move the box right to left and up and down until you see a rainbow of colors above or on the scale. Note: Look at the scale, not at the slit. If the spectrum appears too dim, try covering the top of the box with your hand. When you use the spectroscope outside at night you will have the reverse problem, the spectrum is bright, but you cannot read the scale. A small flash light can be used, but care must be taken to use only enough light to allow you to see the scale.

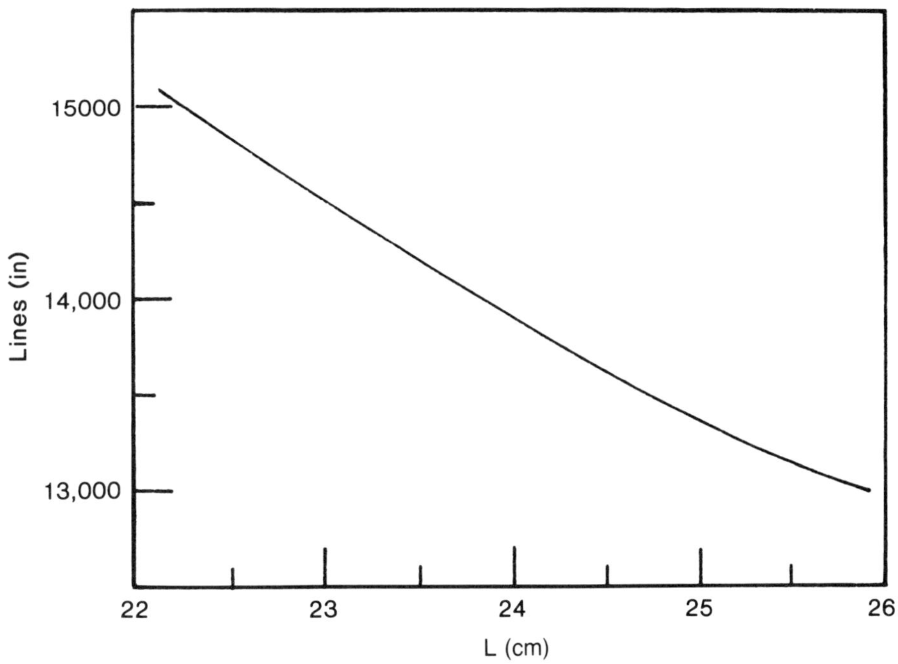

Figure 4

Unit 3.6 Name ______________________________

Date due ____________ Instructor ____________________ Section __________ Seat ____

DO NOT LOOK AT THE SUN WITH THIS INSTRUMENT

1) Use the spectroscope to observe the spectrum from an incandescent light. List the colors that you see.

____

____

____

____

____

____

2) The spectrum has wavelength from ____________ Å to ____________ Å.

3) Use the spectroscope to observe the spectrum from a fluorescent light (try to find one that does not have a diffusing panel over the fluorescent tubes). Compare this spectrum with the spectrum from the incandescent light.

4) Most street lights will be yellow white, light blue, or amber, with the light blue and amber being most common.

Find a light blue street light and look at its spectrum. Measure the wavelengths of the lines that you observe and list their colors.

Wavelength (Å)	Color
____________________	______________
____________________	______________
____________________	______________
____________________	______________
____________________	______________

5) Find an amber street light and observe its spectrum. Describe exactly what you see, giving colors and wavelengths.

6) Instructor assigned sources

A) ______________________________________

B) ______________________________________

C) ______________________________________

D) ______________________________________

E) ______________________________________

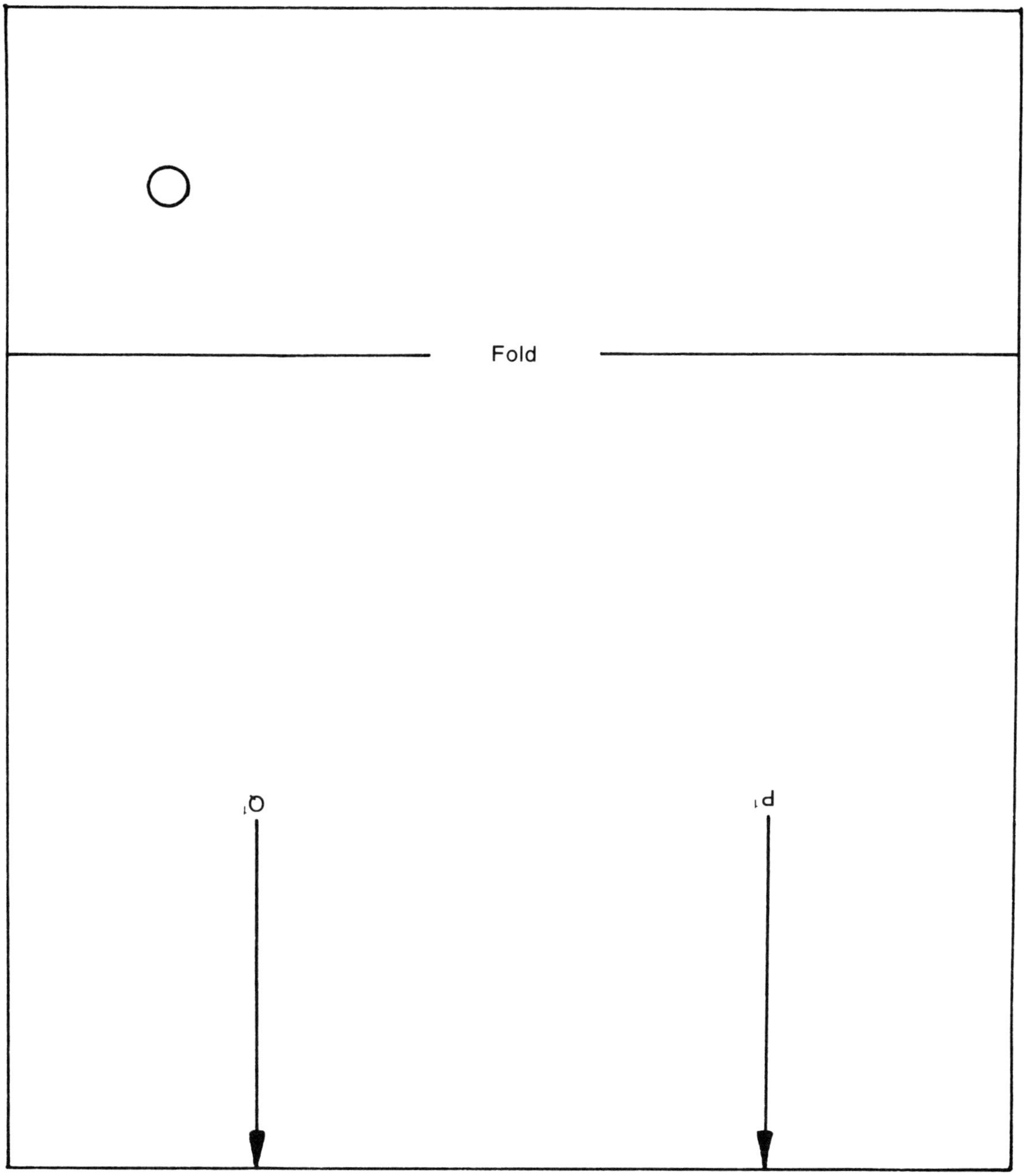

Figure 5

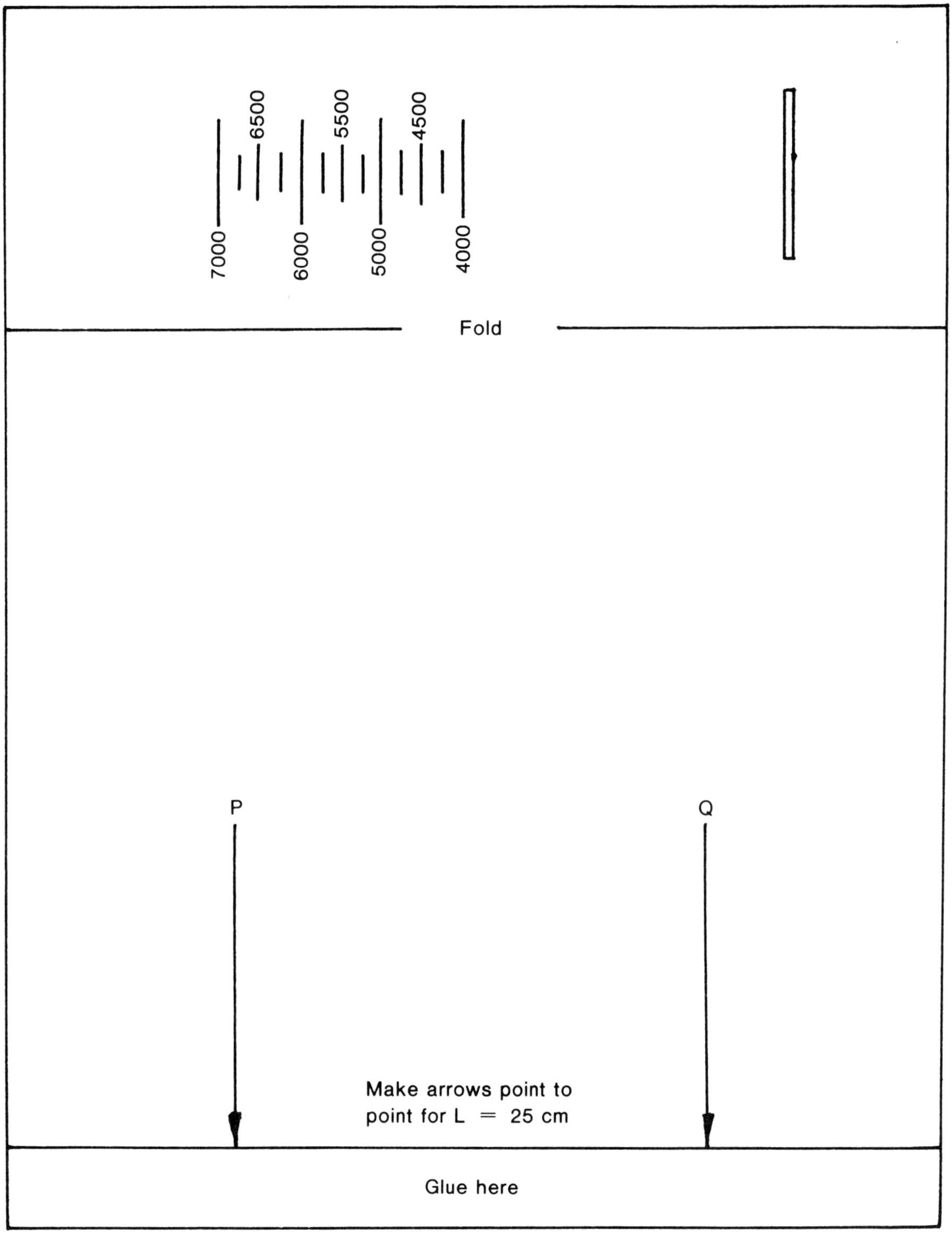
7000
6500
6000
5500
5000
4500
4000
Fold
P
Q
Make arrows point to
point for L = 25 cm
Glue here

Figure 6

4

A CLOSER LOOK THROUGH BINOCULARS AND TELESCOPES

Unit 4.1

Lunar Features

OBJECTIVE To observe the Moon, its phases and features.

INTRODUCTION Our moon is still one of the most fascinating objects to observe. Some of its features you can see with the naked eye, many more through binoculars, and much detail can be seen through astronomical telescopes.

A) NAKED EYE OBSERVATION

For purposes of this part you should look for three things:

1) the phase of the moon,
2) the maria, particularly Mare Crisium. It is almost circular in outline, and easy to recognize on the West limb of the moon. You can always "use" this mare for purposes of orientation. Most telescopes invert the image and exchange the right side with the left side. In the sketches of the moon which you are required to draw, the moon must always have the same orientation.
3) the earth shine. It is easily visible only if the phase of the moon is crescent.

B) OBSERVATION WITH BINOCULARS

In addition to the maria, you can now identify some of the larger craters. If you lie on a lounging chair and keep the binoculars steady, you will be amazed how much detail you can see. Use binoculars frequently, they have also the advantage that you see the true orientation of the moon—as with un-aided eyes.

C) OBSERVATION WITH ASTRONOMICAL TELESCOPES

You will see many interesting details. Keep your eye open for the following features:

1) craters with central peaks, craters with secondary craters on their floors or on their rims,
2) craters with a "ray-system" associated,
3) mountain ranges,
4) rilles.

You will also notice that the same features are not always equally well visible. You have encountered the same situation here on earth: Imagine you look into the Grand Canyon while the Sun is high in the sky. There are hardly any shadows, everything looks flat. You see the Canyon much more "plastic" if there are shadows, i.e. the Sun is relatively low in the sky. The same situation prevails on the lunar surface. At full moon, the Sun's rays strike the lunar surface from straight overhead, there are no shadows (except towards the rim), and the moon is enormously bright. It is difficult to see much detail unless filters are used. The most impressive viewing is during the crescent phases and close to the terminator—the line dividing the sunlit side of the moon from the dark side.

PROCEDURE

You are required to make ______ observations of the moon and to identify by name: 3 craters, one of them with a ray system; 2 maria; 1 mountain range.

When you make the observations, proceed in the following way:

1) Write the date and the time of observation in the appropriate place on the observation sheet, p. 4–3 and 4–5.
2) Draw the phase of the moon, use whatever part of the circle is needed. Do this with the un-aided eye. Indicate Mare Crisium as long as it is visible.
3) Use either binoculars or an astronomical telescope to identify the required features. Write in the appropriate place what instrument you have used. Include pertinent information, for example 7 × 35 binoculars, 8 inch Celestron with 80 × magnification, or whatever is applicable.
4) Make a sketch of some outstanding craters, maria, mountain ranges or whatever other interesting features you see—in the appropriate place of the outlined phase of the moon.
5) Identify by name those features. Use the list of lunar features below. Of course, you are welcome to identify any number of additional features, which are not required. Use your textbook or other available sources, e.g. from the library.
6) Note in "other features" on the observation sheets anything else interesting, for example: an observed rille, other, not listed features, an occultation of a star by the moon, a particularly noticeable earth-shine, etc.
7) You may wish to use more space than is provided by the circle on the observation sheet. For example: You see a large, well defined crater with another, secondary crater on the first crater's rim, and possibly another, tertiary crater sitting on the rim of the secondary crater. You would like to know their names, so you make a large sketch **ON THE BACKSIDE** of the observation sheet. Use that for identification.

LUNAR FEATURES

Craters	Maria	Mountains
Langrenus	Mare Crisium	Caucasus
Aristoteles	M. Foecunditatis	Apennines
Eudoxus	M. Nectaris	Alps
Hercules	M. Tranquillitatis	
Atlas	M. Serenitatis	
Theophilus	M. Imbrium	
Piccolomini	M. Nubium	
Hipparchus	M. Humorum	
Plato	Oceanus Procellarum	
Archimedes		
Copernicus		
Kepler		
Aristarchus		
Eratosthenes		
Tycho		

Unit 4.1 Name ____________________

Date due __________ Instructor ____________________ Section __________ Seat _____

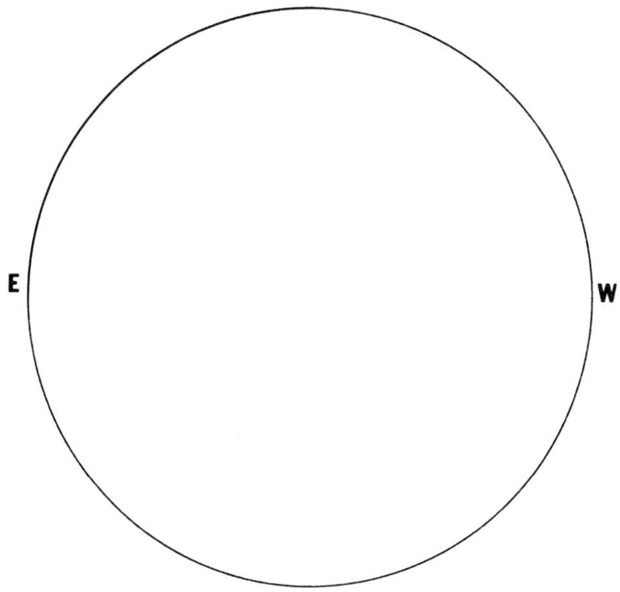

Date observed: __________ Time: __________

Observed through: ____________________

Magnification: ____________________

Identified Features

Craters: ____________________

____

____

Maria: ____________________

____

Mountains: ____________________

other features: ____________________

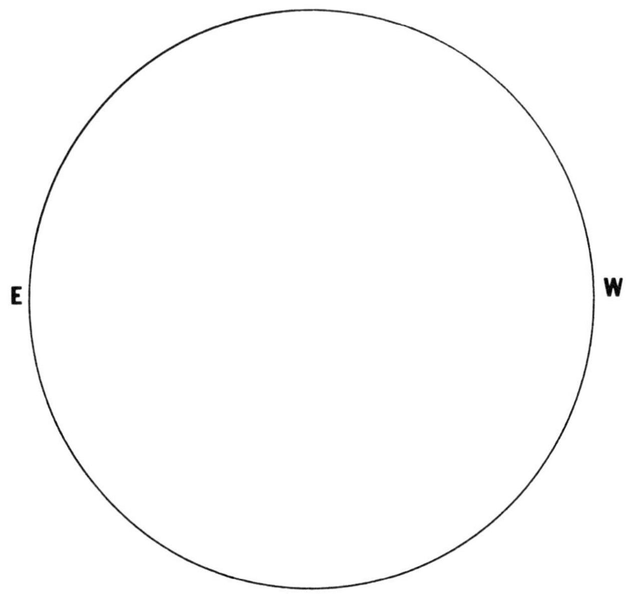

Date observed: __________ Time: __________

Observed through: ____________________

Magnification: ____________________

Identified Features

Craters: ____________________

____

____

Maria: ____________________

____

Mountains: ____________________

other features: ____________________

____

Unit 4.1 Name ____________________

Date due __________ Instructor ____________________ Section __________ Seat ______

E W

Date observed: __________ Time: __________

Observed through: ____________________

Magnification: ____________________

Identified Features

Craters: ____________________

Maria: ____________________

Mountains: ____________________

other features: ____________________

E W

Date observed: __________ Time: __________

Observed through: ____________________

Magnification: ____________________

Identified Features

Craters: ____________________

Maria: ____________________

Mountains: ____________________

other features: ____________________

Unit 4.2
Jupiter and Its Galilean Moons

OBJECTIVE

To observe Jupiter and its Galilean Moons and to note their changing positions.

INTRODUCTION

This unit involves repeated observations of Jupiter. As all planets, Jupiter appears to move slowly through the constellations. It stays always very close to the ecliptic, and is therefore found in one of the Zodiac constellations.

Jupiter itself will appear as a full disk through any astronomical telescope. You should have no difficulty to see also some dark bands across its disk. These bands are parallel to Jupiter's equator. Keep your eyes open for the famous **RED SPOT.** It is not always visible since Jupiter rotates around its axis about every ten hours and the red spot takes part in this rotation. Occasionally you may also see the shadow of one of its Galilean Moons on its disk. It appears as a very small dark dot.

Jupiter has at least thirteen moons. Four of them are easily visible through a telescope. They were discovered by Galileo Galilei, hence: "Galilean" moons. Their names are: Io, Europa, Ganymede, and Callisto, listed from innermost to outermost position. They revolve around Jupiter in nearly circular orbits in a plane around Jupiter's equator. Since we look very nearly along Jupiter's equatorial plane, the moons therefore seem to move back and forth along a line through Jupiter's disk.

PROCEDURE

Before you go out for observations of Jupiter, make sure that you know where in the sky it is located, i.e. in which Zodiac constellation. Then:

1) Draw an outline of the constellation in the big square on p. 4–9 for the beginning of your observation series. You may have to draw two constellations if Jupiter happens to be located right between them. Mark Jupiter's position within that constellation by an (x) and note the date. You are welcome to use the constellation maps as help for your drawings.
2) Look at Jupiter through an astronomical telescope. Be sure that you can see at least an indication of the dark bands. That fixes the direction of Jupiter's equatorial plane. Then look for the moons in the extension of the equatorial plane. You may not see all four of them, one or more could be either in front or behind Jupiter's disk. Before you draw the required sketch of the moons' positions, it is necessary to spend some time to consider the orientation of Jupiter as you see it through the telescope:

   An astronomical telescope inverts the image and changes sides (left and right) as well. In addition, a star diagonal is used very often for the sake of convenience. If you rotate the star diagonal, the image rotates. It is therefore necessary to consider carefully Jupiter's orientation. An easy way out is the following: If a binocular is available, look through it. You see the same orientation as you see with your unaided eye. You will very probably see at least the brightest of the four Galilean moons through the binocular. Mentally note its alignment with the planet. Then look through the astronomical telescope. If you own an astronomical telescope yourself, you can always use it to look at an object on earth. Notice how the telescope (with star diagonal) changes the orientation of the image. Make some sketches and apply them to the orientation of Jupiter. Otherwise, ask a laboratory assistant for information.

You **MUST** draw all the successive sketches of Jupiter's moons **IN THE SAME ORIENTATION.** Make a note, the first time you observe them, about the alignment of the star diagonal. Use the same alignment for successive observations.

3) Of equal importance as the orientation of Jupiter's image is the magnification you use for observation. It should be small enough so that you can see all of Jupiter's Galilean moons. Use **THE SAME MAGNIFICATION** for all successive observations, and note in the appropriate place on p. 4–9.
4) Finally, draw Jupiter's moons in the appropriate orientation and spacing in the rectangular fields on p. 4–9.

   Include the date and the time of observations.

5) Make at least ______ observations of the moons within a time period of no more than two weeks.

   Very often it is easily possible to make more than one observation on any given night. You may, for example, observe one of the moons very close to Jupiter's disk. Make a sketch. A short while later you may notice that this moon has either moved noticeably farther away from the disk or moved closer. If the latter is the case, make it a point to observe the moon vanishing. It will either move behind Jupiter's disk, in which case it may be worth your while to wait until it reappears on the other side. That will be a fine observation for another sketch. In case the moon moves across the disk in front, you may want to see its shadow on the disk, and possibly also wait until it reappears. Information about the disappearance of Jupiter's moons will be available from
   a) "Sky and Telescope" magazine or from
   b) your instructor or the laboratory assistants.
6) If you have made about three or four observations, it should be possible to identify the four Galilean moons by name. Use the information in the table below. Identify the moons by their Roman numerals in the drawings on p. 4–9.

Name	distance from Jupiter in km	sidereal period in days
I Io	420,000	1.8
II Europa	670,000	3.6
III Ganymede	1,100,000	7.2
IV Callisto	1,900,000	16.7

7) When you are finished with the observations of Jupiter's moons, look again at Jupiter's position in the constellation. If Jupiter has moved noticeably during this time interval, make another sketch of Jupiter's position in the second (lower) large square, and date.

Unit 4.2 Name ____________________

Date due ______________ Instructor ____________________ Section __________ Seat ____

Location of Jupiter relative to the stellar background. Date: _______	**Jupiter and Its Galilean Moons** Telescope and Magnification: __________
	⊖ Date: Time:
	⊖ Date: Time:
	⊖ Date: Time:
Location of Jupiter relative to the stellar background. Date: _______	⊖ Date: Time:
	⊖ Date: Time:
	⊖ Date: Time:

Unit 4.3

Mercury, Venus, Mars, Saturn

OBJECTIVE To observe and make sketches of these planets

INTRODUCTION All of these planets can be seen with the naked eye, provided they are in a favorable position. Venus and Saturn can be a spectacular sight even through a small astronomical telescope.

MERCURY

It is visible either shortly before sunrise or shortly after sunset. It is seen as a small disk—and shows phases. No other features are observable.

VENUS

No other planet appears at certain times as bright as Venus. It can be seen for a maximum time of about 3 hours as an "evening star" after sunset, or for about 3 hours maximum time as a "morning Star" before sunrise. At its brightest, Venus can even be seen after the sun has risen.

It shows phases as the Moon and Mercury. At its crescent phase, its angular diameter is quite large and Venus presents a spectacular view through an astronomical telescope. No other features of Venus are observable.

MARS

It looks always reddish, and is therefore easy to recognize among the planets. Under good viewing conditions you may be able to see a polar icecap and some dark markings on its otherwise reddish surface. Due to its small size, Mars never appears as such a large disk as do Jupiter and Saturn.

SATURN

It presents a large disk, whose oblateness can be seen under favorable conditions. Most spectacular, of course, are its rings. With an 8 inch telescope, Cassini's division can easily be seen, provided that the plane of the rings is much inclined to our line of sight. You may also see some dark bands across the disk. The bands as well as the rings are parallel to Saturn's equator.

The inclination of Saturn's rings will change with time. As we approach the time when we look along Saturn's equatorial plane, the rings will become invisible for a while (they are extremely thin).

Some of Saturn's moons can also be seen. For most of the time, however, they are not "in line" as are Jupiter's moons. It is more difficult to judge which bright spots are Saturn's moons and which are stars. The exception is Saturn's largest moon, Titan, which is relatively easy to follow for successive nights of observations.

PROCEDURE

1) Locate the planet assigned for observation in the sky. Mark the planet's position within that constellation by an (x), and note the date, in the large rectangle on p. 4–13 and 4–15.

2) Observe the planet through a telescope, note the kind of telescope, its magnification, and the date of observation to the right of the circle.
3) Make a sketch of the planet as you see it within the circle (the field of view of the telescope).

4) Additional instructions: ______________________________

____

____

____

Unit 4.3 Name ______________________________

Date due ______________ Instructor ______________________ Section __________ Seat ____

Planet: ______________: Its location within a constellation, or relative to sun and horizon

Date: ______________

Telescope: ______________

Magnification: ______________

Date: ______________

Planet: ______________: Its location within a constellation, or relative to sun and horizon

Date: ______________

Telescope: ______________

Magnification: ______________

Date: ______________

Planet: ______________: Its location within a constellation

Date: ______________

Telescope: ______________

Magnification: ______________

Date: ______________

Planet: ______________: Its location within a constellation

Date: ______________

Telescope: ______________

Magnification: ______________

Date: ______________

Unit 4.3 Name ____________________

Date due ____________ Instructor ____________________ Section ____________ Seat ______

Planet: ____________ : Its location within a constellation, or relative to sun and horizon

Date: ____________

Telescope: ____________

Magnification: ____________

Date: ____________

Planet: ____________ : Its location within a constellation, or relative to sun and horizon

Date: ____________

Telescope: ____________

Magnification: ____________

Date: ____________

Planet: ____________ : Its location within a constellation

Date: ____________

Telescope: ____________

Magnification: ____________

Date: ____________

Planet: ____________ : Its location within a constellation

Date: ____________

Telescope: ____________

Magnification: ____________

Date: ____________

Unit 4.4

Double Stars, Open and Globular Clusters, Nebulae, and Galaxies

OBJECTIVE To locate and observe some of these objects.

INTRODUCTION To the casual observer there are nothing but single stars in the night sky. Some people may have seen a cluster of stars, the Pleiades or seven sisters, and—if the sky was particularly dark and clear—the band of the Milky Way.

Yet the sky abounds with interesting and beautiful objects. You do not even need an astronomical telescope. It is amazing how many interesting objects can be seen through binoculars.

The tables below contain some of the most interesting and easily observable objects. There are, of course, very many others, and you can make your own prefered list from a variety of sources. The objects listed below were selected because a) they are particularly interesting or beautiful, and b) they are easy to find and to observe. No telescope larger than 2 or 3 inches diameter is necessary. Objects in circumpolar constellations are listed first, then from spring to winter constellations.

TABLE 1: Double Stars

object #	Constellation	Name, if any	app. visual magnitude of components m	distance of separation in sec of arc
1)	α in Ursa Minor	Polaris	2; 9	18
2)	ζ in Ursa Major	Mizar	2.4; 4	14
		Mizar/Alcor	2; 4	>600 (12 min of arc)
3)	δ in Cepheus		3.7–4.4; 7.5	41
4)	α in Canes Venatici	Cor Caroli	3.2; 5.7	20
5)	γ in Leo		2.3; 3.5	5
6)	β in Cygnus	Albireo	3; 5	35
7)	ϵ in Lyra		4.2; 5.5	150
			ϵ_1 5; 6	3
			ϵ_2 5; 5.5	2
8)	α in Hercules		3.5; 5.5	5
9)	β in Scorpius		3; 5; 5	1–14
10)	γ in Andromeda		2; 5	10
11)	α in Capricornus		4; 4	375
			α_1 4; 9	45
			α_2 3.8; 9.5	154
12)	α in Gemini	Castor	2; 3	5
13)	β in Monoceros		5; 5; 6	7–10

14) ______________________________

15) ______________________________

TABLE 1 (Continued)

object #	Constellation	Name, if any	app. visual magnitude of components m	distance of separation in sec of arc
16)				
17)				
18)				
19)				

All Messier objects are listed in table 5 in the appendix, p. 9–24. Some of the brightest and best visible are listed here, together with a few objects, not in the Messier table.

TABLE 2: Open and Globular Clusters, Nebulae, Galaxies

20) M 81, M 82	in Ursa Major	these two spiral galaxies are only 0.5° apart and may be seen in the same field of view. M 81 is very elongated. m: 8 and 9.
21) M 52	in Cassiopeia	open cluster, m: 7.3
22) NGC 663	in Cassiopeia	open cluster
23) NGC 7789	in Cassiopeia	open cluster
24) M 44	in Cancer	Praesepe or Bee-Hive open cluster, visible with naked eye, best view through binoculars.
25) M 3	in Canes Venatici	globular cluster, m: 6.3
26) M 51	in Canes Venatici	the famous Whirlpool galaxy, m: 8
27) NGC 5139	in Centaurus	large globular cluster, not visible at latitudes much higher than 35° N.
28) M 13	in Hercules	the famous Hercules cluster, very bright, m: 5.7. Together with M 22 the finest globular cluster.
29) M 57	in Lyra	the famous ring nebula, planetary nebula, you should see this object as a ring, a smoke-ring or a dough-nut. The center star is visible only through large telescopes. m: 9.3
30) M 27	in Vulpecula	the Dumbbell nebula. Another planetary nebula. On clear, dark nights extemely impressive, looks almost three-dimensional. m: 7.6
31) M 8	in Sagittarius	Lagoon nebula, diffuse nebula, visible to naked eye.
32) M 17	in Sagittarius	Horseshoe nebula, diffuse nebula.
33) M 20	in Sagittarius	Trifid nebula, diffuse nebula.
34) M 22	in Sagittarius	globular cluster, one of the finest, m: 5.9
35) M 24	in Sagittarius	open cluster, m: 4.6
36) M 4	in Scorpius	globular cluster, very compact, m: 6.5

TABLE 2 (Continued)

37) M 6, M 7	in Scorpius	open clusters, in the same field with wide-angle binoculars, m: 5.5
38) M 80	in Scorpius	globular cluster, m: 8
39) M 31	in Andromeda	the great Andromeda galaxy, the only spiral galaxy visible to the naked eye, 2 mill. lightyears distant, m: 4.8
40) M 32	in Andromeda	an elliptical galaxy, companion to M 31, m: 9
41) M 34	in Perseus	open cluster, m: 5.5
42) NGC 869 and 884	in Perseus	the famous double cluster, use very low power to see both in the same field of view.
43) M 33	in Triangulum	spiral galaxy, m: 6
44) M 2	in Aquarius	globular cluster, m: 6
45) M 15	in Pegasus	globular cluster, m: 5.2
46) M 37	in Auriga	open cluster, m: 6
47) M 35	in Gemini	open cluster, m: 5.3
48) M 1	in Taurus	Crab nebula, remains of the 1054 supernova explosion, faint cloud, m: 8.4
49) M 45	in Taurus	Pleiades or Seven Sisters, open cluster. Best view through binoculars.
50) M 41	in Canis Major	open cluster, m: 4.5
51) M 42	in Orion	diffuse nebula, the Great Orion Nebula, in the sword. Brightest of the diffuse nebulae. Look for "Trapezium", a quadruple star system in the nebula.

52) ______________________________

53) ______________________________

54) ______________________________

PROCEDURE

Objects, required for observations and drawings:

Object #'s: 49), 51) and Jupiter

1) Write the object #, its name and/or designation, and its nature on the lines within the rectangles on pages 4–21, 4–23, 4–25.
2) Make a sketch of the constellation or the asterism in which the object is located. You are welcome to use the star maps for this purpose.
3) Indicate by (x) or other appropriate means the location of the object within the constellation.
4) Observe the object through a telescope and make a sketch of the object. If you have never seen an open cluster, a globular cluster, a nebula, or a galaxy before, it may be very much worthwhile to look at these different objects through a scope before you attempt to make a drawing. For example, it is impossible to make an accurate plot of all the stars which form an open (or galactic) cluster. Looking at a globular cluster and comparing it with an open

cluster, you will see the fundamental structural difference. This will help you to make the drawing of either one.

5) Fill in the required information for the telescopic observation, including the date of observation.
6) Some objects are best seen through binoculars, e.g. the Pleiades. Write "binocular" on the line: Telescope used:
7) For the observation and drawing of Mizar/Alcor: Look through a binocular or the finderscope of the astronomical telescope first. Make a drawing in the small circle. Label Mizar and Alcor. Then look through the astronomical telescope and make the appropriate drawing in the large circle, also label Mizar and Alcor. Do this on p. 4–21.
8) When you observe double stars, you will note that very often one of the components is brighter than the other. Use larger dots for brighter stars. Also, the colors of the components may be noticeably different from each other. If observed, note the colors.
9) ______________________________

____

____

10) ______________________________

____

____

Unit 4.4 Name ______

Date due ______ Instructor ______ Section ______ Seat ______

Object #: 2), Name or Designation: Mizar/Alcor

Nature of object: multiple star system

Location within constellation:

Telescope used: ______

Magnification: ______X

Date: ______

through binoculars or finderscope

Object #: ___, Name or Designation: ______

Nature of object: ______

Location within constellation:

Telescope used: ______

Magnification: ______X

Date: ______

Object #: ___, Name or Designation: ______

Nature of object: ______

Location within constellation:

Telescope used: ______

Magnification: ______X

Date: ______

Unit 4.4 Name ______

Date due ______ Instructor ______ Section ______ Seat ______

Object #: ____, Name or Designation: ______

Nature of object: ______

Location within constellation:

Telescope used: ______

Magnification: ______ X

Date: ____

Object #: ____, Name or Designation: ______

Nature of object: ______

Location within constellation:

Telescope used: ______

Magnification: ______ X

Date: ____

Object #: ____, Name or Designation: ______

Nature of object: ______

Location within constellation:

Telescope used: ______

Magnification: ______ X

Date: ____

5

A CLOSER LOOK WITH OTHER TOOLS

Unit 5.1

The Altitude—Azimuth Coordinate System

OBJECTIVE To become familiar with the Alt-Azimuth coordinate system.

INTRODUCTION This system is a very simple one and very often used. You use its azimuth part every day you drive a car. The highways are designed as E 66 or W 24 or N 101 or S 101, etc. You use the points of the compass to denote a direction. An alt-azimuth system is used by a tripod, for example. However, the photographer usually does not care to read any angles. The camera is moved horizontally about an axis—it moves through azimuth angles; it also can be moved perpendicularly to this horizontal plane—through an altitude angle.

The least expensive astronomical telescope mount is a simple alt-azimuth mount. The telescope rests in a fork and can be moved up and down, pointing towards the earth or up into the sky. The fork with the telescope can be moved in a horizontal plane, the azimuth plane. If you want to observe a star for longer periods of time, you would then have to change the azimuth as well as the altitude of the telescope constantly, because the stars seem to move in circular arcs.

The alt-azimuth system consists of a reference plane, the horizon plane and its pole, the zenith. The azimuth is measured along the horizon, starting with 0° at the Northpoint, in an easterly direction, with East (E) 90°, South (S) 180°, West (W) 270° and finally North (N) with 360° or 0°. The altitude is measured from the horizon with 0° upwards to the zenith with 90°. In addition to the four cardinal points of the compass, N, E, S, and W the half-way direction between any of them is commonly used, namely NE, SE, SW, and NW. A still finer division of the azimuth scale denotes the halfway mark between N and NE as NNE—which means north of northeast. ENE means east of NE. Similarly for the other directions. However, the use of this terminology should be restricted in the following way: If an object is E of an observer, it has an azimuth of 90°; if it is NE of an observer, it has an azimuth of 45°; if it is ENE of an observer, it has an azimuth of 67.5°.

If you do not wish to denote the direction to an object too accurately and merely wish to say that it is somewhere to the north of East, then you should say so in words, namely North of East. You should **NOT** say: NE because that specifies a well defined direction, namely 45° azimuth. Study Fig. 1 for angles and points of the compass. In astronomy the alt-azimuth system is very often used to describe events which change relatively fast. For example, locations of comets, conjunction of planets, satellite orbits, etc.

For some of the projects in this manual you need to employ the alt-azimuth system. That is simple enough except for the determination of the 0° direction, i.e.

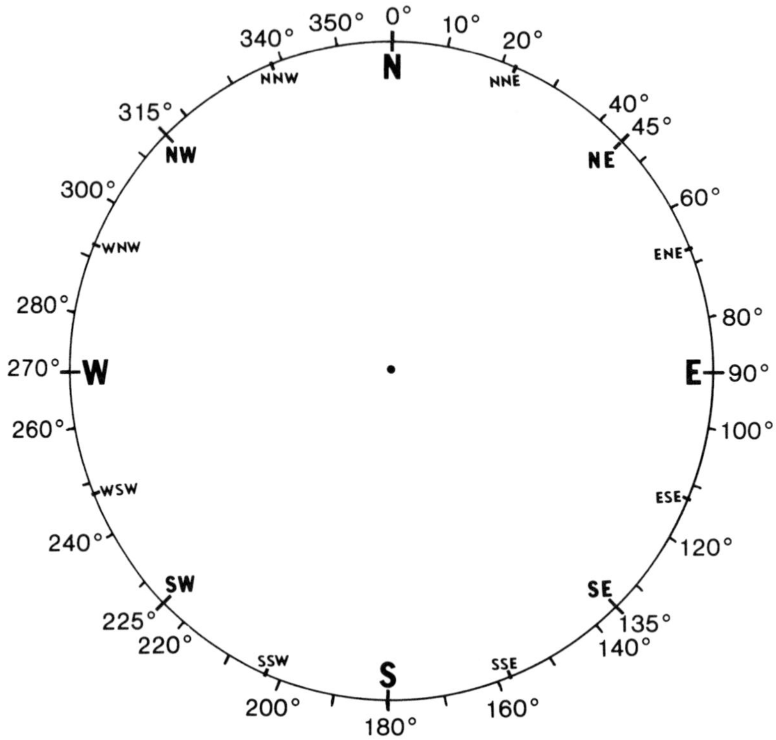

Figure 1

North. There is a number of ways how you can do that, a few of them are described below:

DETERMINATION OF NORTH (0° AZIMUTH)

1) At night.
   Look at Polaris. A line straight down from it goes close to the North-point on the horizon. With a piece of string, a rod or similar, simple devices you can easily mark the North direction. This method is generally sufficient if the required accuracy in angle measurements is not more than ±1°. True North is, however, the direction towards the North Celestial Pole (NCP). Polaris is at present 0.8° off. When Cassiopeia or the handle of the big dipper are either above or below Polaris, then NCP, Polaris, and the Northpoint on the horizon are in line. If Cassiopeia and the handle of the big dipper are to the side (E or W) of Polaris, then the NCP is either 0.8° to the W or E of Polaris, respectively. Cassiopeia (specifically epsilon Cassiopeiae) and Polaris are on the same side of the NCP. Consult the map of circumpolar stars, p. 2–23.
2) During the day.
   The sun goes through your meridian at High Noon. That means that the sun is highest in the sky when it is exactly south. High Noon, however, is not the same as 12:00 noon on your clock. Depending on your geographical longitude, your clock shows either High Noon to be before, on, or after 12:00 noon. Consult your instructor for the time of High Noon or consult Unit 5.4. Assume that at a certain location High Noon is at 12 hr 28 min pm Standard Time on your clock. The shadow of any straight object, perpendicular to the horizon, is along the N-S direction at 12 hr 28 min pm Standard Time.

3) Topographic maps of large enough scale are very accurate. If your location for observation is close to identifiable lines on the map, you may take the N-S direction from the map.
4) You may use a magnetic compass. However, the N-direction on the compass is magnetic North. To deduce geographic North from that you must know the magnetic declination. Consult your instructor, if necessary.

EXERCISES

Answer the questions on p. 5–5.

Unit 5.1 Name ______________________________

Date due ______________ Instructor ______________________ Section ____________ Seat _____

Answer in empty spaces:

Description of event	Points of Compass use E, NE, NNE, etc.	Azimuth in degrees	Altitude in degrees
1) The sun rises exactly East	E		
2) The sun rises exactly ESE	ESE		
3) The sun sets 22.5° north of west			
4) The moon is exactly south and halfway between horizon and zenith			
5) Jupiter goes through your meridian and is straight overhead			
6) The moon occults Venus 45° south of west and ⅓ the distance between horizon and the zenith			
7) A satellite "rises" at 8 hr 00 min pm exactly at WNW ______	WNW		
2 minutes later it passes through the zenith. Where will it "set" at 8 hr 04 min			

Give the corresponding azimuth values for the following points of the compass:

1) NE ____________ 7) NW ____________

2) NNE ____________ 8) NNW ____________

3) ENE ____________ 9) WNW ____________

4) SE ____________ 10) SW ____________

5) ESE ____________ 11) SSW ____________

6) SSE ____________ 12) WSW ____________

Unit 5.2

Time and/or Azimuth of Sunrise and/or Sunset throughout the Year

OBJECTIVE

To make a graph of the changing times and azimuths of sunrise and/or sunset versus the days throughout the year.

INTRODUCTION

The winter solstice is the shortest day: The sun rises latest in the morning and farthest to the south, it sets earliest in the evening and farthest to the south. At High Noon, it is lowest in the sky. After this day the sun rises earlier and farther north, is higher in the sky at High Noon, and sets later and farther to the north until the summer solstice. Then the process is reversed back to the winter solstice.

Consider a certain day of the year, for example January 1: The sun will rise at a certain time at a certain azimuth angle. What will the time and the angle be the next day, January 2? You expect the time of sunrise to be a little earlier and the azimuth angle a little smaller, that is farther north. How much? Now consider another couple of days, some time later, say April 1 and April 2. The sun rises much earlier than in January and much farther north, of course. Will the difference in time and the difference in angle between April 1 and April 2 be the same as that in January?

It is worthwhile to investigate these questions by plotting the times and angles of sunrise or sunset. In order to extract all possible information from the plot, it would be necessary to determine times and azimuth angles for at least one quarter year, including one of the equinoxes or solstices at the beginning (or end) of the period for the plot. This may not be possible for any particular course you take but you may wish to do that on your own, in addition to any assignments. Consult your instructor.

PROCEDURE

A) Times of Sunrise and/or Sunset

1) The data for the plot can be collected in the following ways:
   a) From weather data, published e.g. in the daily newspaper.
   b) From printed tables, e.g. the American Ephemeris and Nautical Almanac.
   c) From your own measurements.

   Use the method as specified by your instructor.
2) Consult the graph on p. 5–9. According to the period which will be specified by your instructor, label the dates on the abscissa, from the begin of the project to its end. Your instructor will specify the range of times for rising or setting which you then label on the ordinate. Note the 6 am or 6 pm line at the center of the page. In general, the sun will rise later than 6 am and will set earlier than 6 pm between the autumnal and the vernal equinox. Your instructor will also specify the kind of time to be used, that is: Standard Time or local solar time, or possibly Daylight Saving Time.
3) Plot the times of rise and/or set of the sun versus the date. A plot for every other day is quite sufficient if you use published data. If you measure the times yourself, once a week is adequate.
4) When the project is finished, connect the plotted points by a line.

B) Azimuth angles of Sunrise and/or Sunset

1) The data for the plot can be collected in the following ways:
   a) From planetarium demonstrations.
   b) From celestial globes which contain a movable sun and a graduated horizon plane.
   c) From your own measurements. The measurements can be performed by using an azimuth scale and a pin, sufficiently long, at the center of the azimuth scale. The scale itself must sit on a level platform and the pin must be vertically affixed to the scale. Read the azimuth once or twice a week at sunrise or a certain time after sunrise, but always the same time interval after sunrise. You may use the starfinder for altitude, azimuth measurements.
   d) If you have made your star-tracker and have read Unit. 2.1, then you may know of still another method to collect data. Think about it!

   Use the method as specified by your instructor.
2) Consult the graph on p. 5–11. Your instructor will specify the ranges to be used, both for the abscissa and the ordinate. Note the center line in the graph for an azimuth of either 90° (East) or 270° (West). Label the abscissa and the ordinate as instructed.
3) Plot the azimuth angle versus date, once or twice a week for the specified period.
4) When the plot is finished, connect all points by a line.

QUESTIONS

1) On which date of the year does the sun rise earliest in the day?
2) On which date of the year does the sun set latest in the day?
3) On which dates of the year is the change in rising or setting times from one day to the next one smallest? largest?
4) On which date of the year does the sun rise farthest north? farthest south?
5) On which dates of the year does the sun rise or set about E or W, respectively?

Unit 5.2 Name ____________________

Date due ____________ Instructor ____________________ Section ____________ Seat ______

Exercise A

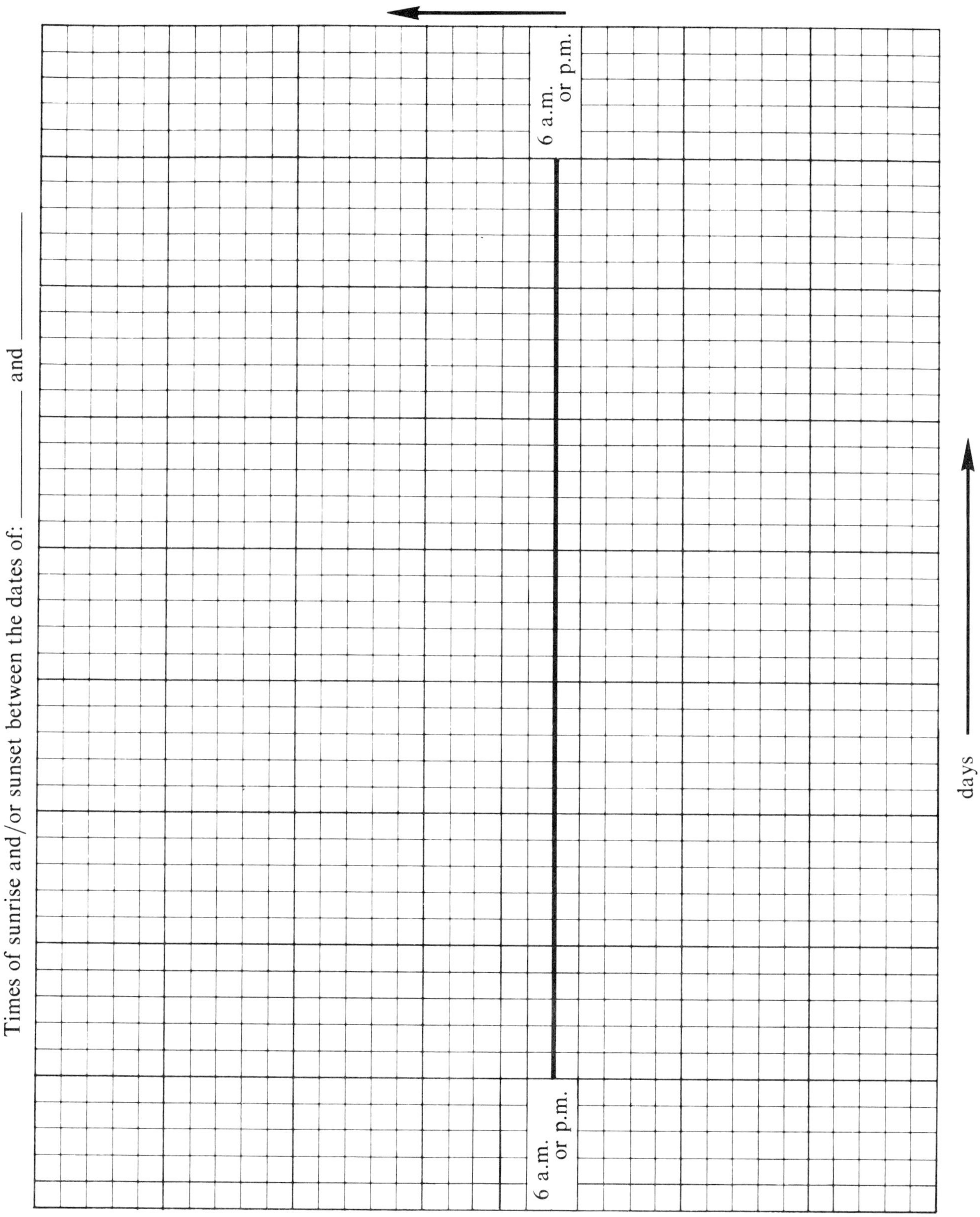

Unit 5.2 Name ______________________________

Date due ______________ Instructor ______________________ Section __________ Seat _____

Exercise B

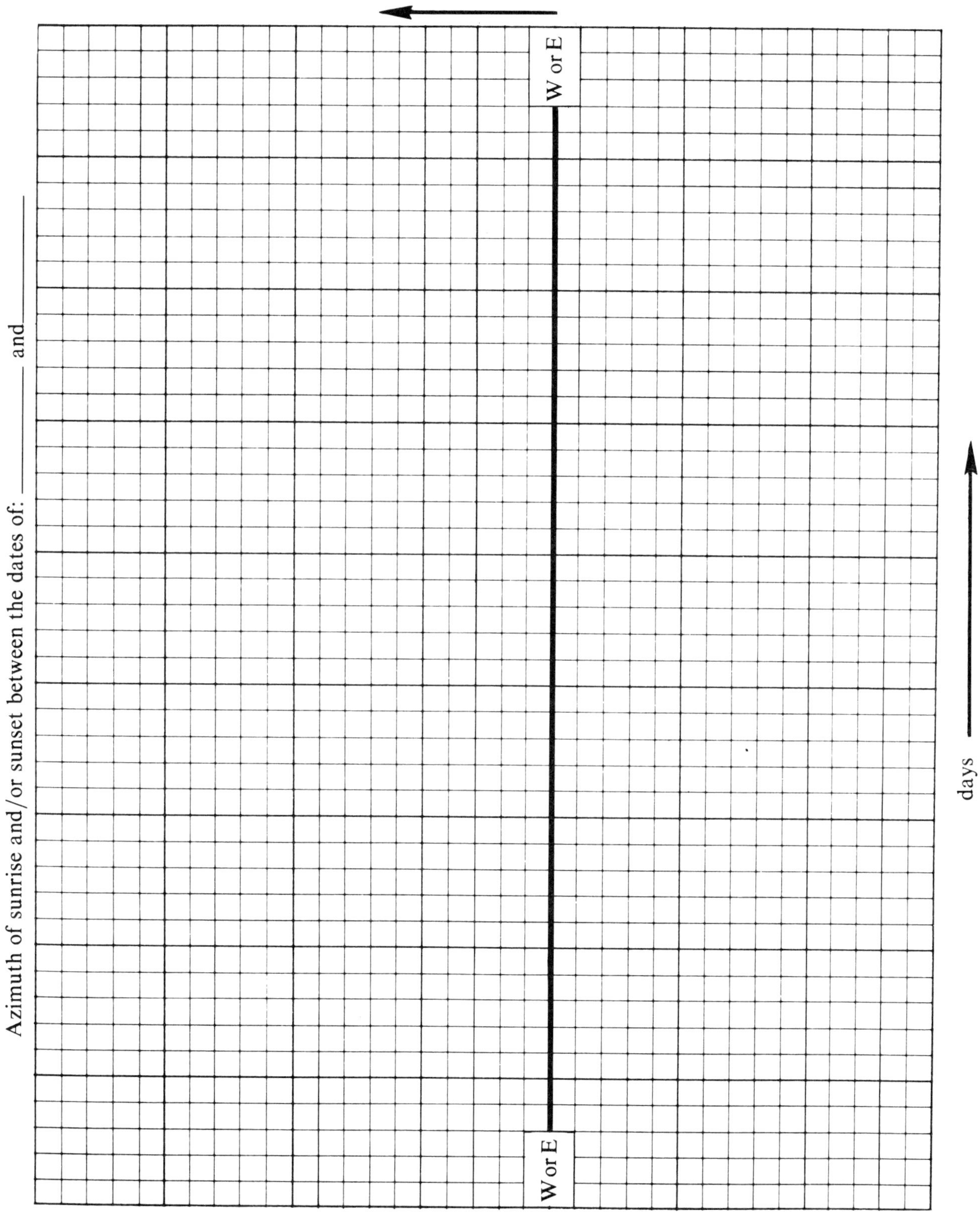

Unit 5.3

The Altitude and Azimuth of the Sun

OBJECTIVE

To measure the Sun's altitude at a certain azimuth and to plot its altitude versus time.

INTRODUCTION

During any given day of the year the Sun rises somewhere along the eastern horizon, increases in elevation above the horizon (altitude), reaches its highest altitude at High Noon, then sets somewhere along the western horizon.

At High Noon the Sun goes through your meridian and is exactly south. High Noon, however, does not necessarily coincide with 12 hr 00 min on your clock. Your clock shows either Standard Time or Daylight Savings Time. See Unit 5.4 for these different times, or consult your instructor.

You are, of course, aware of the fact that the Sun is generally higher in the sky during the summer months than during the winter. The Sun's altitude therefore depends on the time of the day as well as on the day of the year.

APPARATUS

The starfinder of Unit 3.3 or 3.4, or a protractor for altitude measurement. The use of the starfinder for altitude-azimuth measurement is described on p. 3–25.

A protractor can easily be converted to measure both azimuth and altitude. Drill a hole through the center of the protractor to accommodate a thin rod. Have a short piece of string ready, with attached weight, to slip over the rod. Finally, you need a clamp and a support.

In Fig. 1 the protractor lies on a level surface with the rod straight up. If the time is High Noon, then the shadow of the rod is exactly N-S. If the protractor is turned so that the shadow goes through the 0° mark, then the azimuth scale is properly aligned.

In Fig. 2 a straw is attached to the protractor which is held by a clamp to a simple support. The string with attached weight is slipped over the rod which allows you to read an angle. Suppose the straw would be pointing to the Sun, the angle you read in this figure is 66°. The altitude of the Sun then is $90° - 66° = 24°$. Project the image of the Sun. **NEVER LOOK AT THE SUN DIRECTLY.**

PROCEDURE

1) Determine the N-S direction and mark on the surface (level) you are going to use for this project. See Unit 5.1 for ways to find N-S.
2) Set up your starfinder or the protractor according to instructions for altitude, azimuth measurements. You do not need to measure the Sun's altitude at High Noon. Any time between 9 am and 3 pm will do. However, you **MUST MEASURE THE ALTITUDE ALWAYS AT THE SAME TIME OF THE DAY.**
3) At your chosen time, project the image of the Sun on a piece of paper. The correct projection is shown in Fig. 4 below. Two inaccurate projections are in Fig. 3 and Fig. 5 at either side of the correct one.

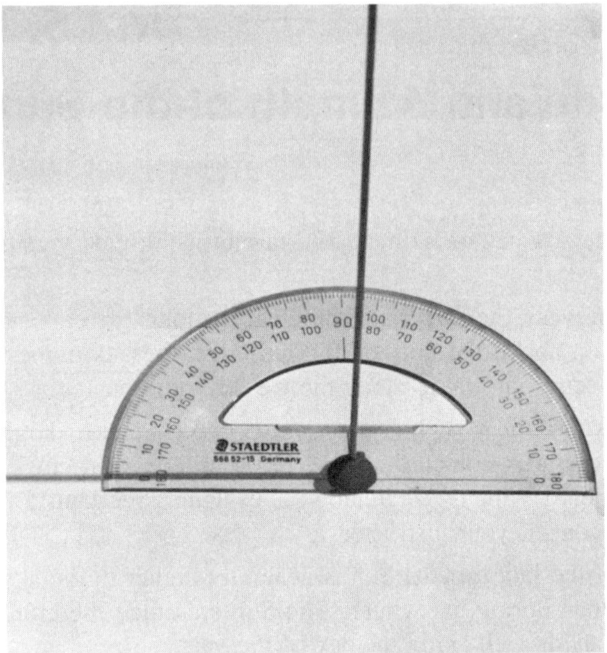

Figure 1

Figure 2

Figure 3. Incorrect

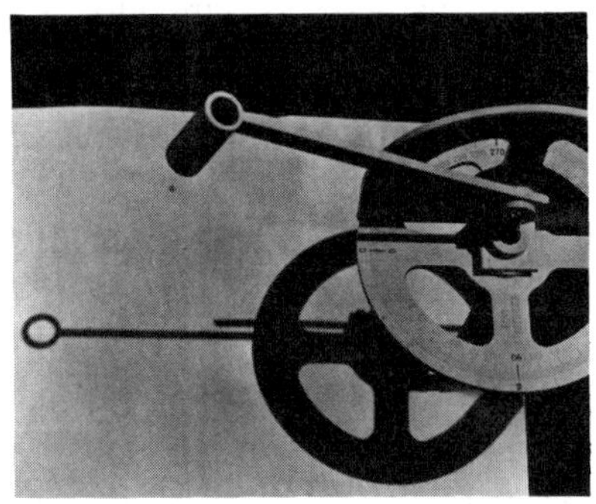

Figure 4. Correct

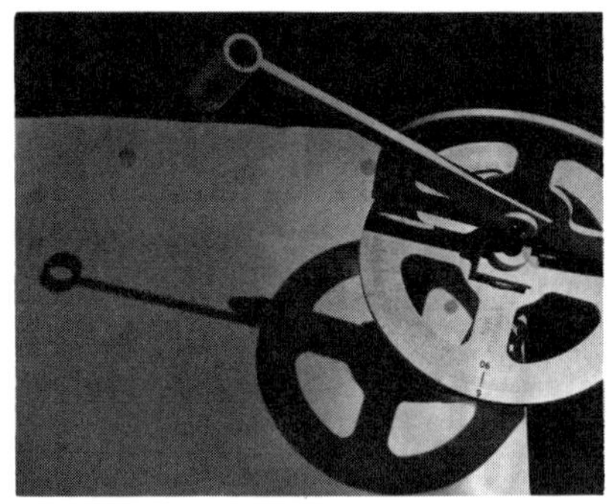

Figure 5. Incorrect

4) Read the Declination or Altitude scale and record in the table on p. 5–15 together with the date.
5) Read the RA scale on the starfinder and convert to azimuth value or read the azimuth scale of the protractor and record on p. 5–15. Remember that the azimuth must be the same throughout the project.

6) The minimum number of required measurements is: ____________ . The measurements must be at least 2 days apart.
7) The first measurement should be as close as possible to the beginning day of the project, the last measurement as close as possible to the "Date due" of the project.
8) Fill in all required information on p. 5–15. Plot the measurements, altitude versus date. Label the date scale (abscissa) as specified by your instructor.
9) If you wish to continue this project on your own for a longer time, use the graph sheets at the end of the manual for an extension of your date scale.

Unit 5.3 Name ______________________________

Date due ______________ Instructor ______________________ Section __________ Seat ____

The measurements were made at what time of day: __________ At this time the sun's azimuth is: _____° This azimuth corresponds to:

_______° E of South
W of South

circle the right compass direction.

The total change in the sun's altitude between your earliest and latest measurement was _____°

earliest date: _____

latest date: _____

Data Table

# date	altitude
1	
2	
3	
4	
5	
6	
7	
8	
9	
10	
11	
12	
13	
14	
15	
16	

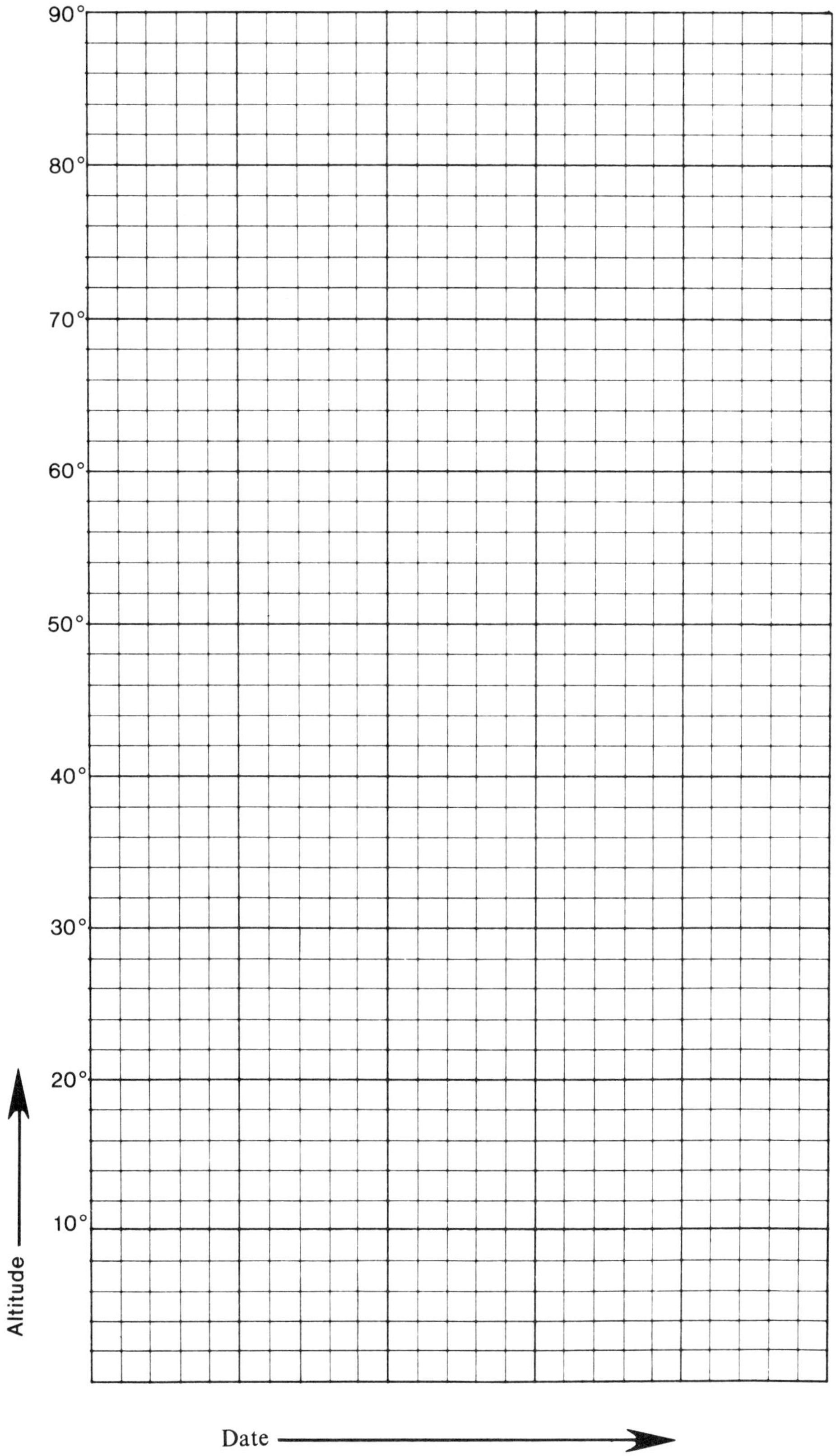

Unit 5.4

Earth's Geographical Coordinates: Latitude and Longitude

OBJECTIVE

To become familiar with an equatorial coordinate system, and to measure your own latitude and longitude.

THEORY

In every-day life, the earth appears as a flat plane to our senses. In order to designate a certain location, we usually are satisfied to specify a direction and a distance. For example, city X is in a north-westerly direction, 12 mi from here.

YOU are at the center of your own coordinate system, as in the Altitude-Azimuth system of Unit 5.1. It would be difficult for different people, living all over our planet, to correlate their own individual coordinate systems with each other. The earth as a planet is **NOT** a flat plane. A coordinate system was introduced, designating locations on the earth's surface, which was identical for everyone, the Latitude-Longitude system.

It consists of a reference plane—the **EQUATORIAL** plane, and its poles, the N- and S-poles. The equator plane is the fundamental **GREAT CIRCLE,** its center is the center of the earth, its plane is perpendicular to the axis of the earth.

At right angles to this fundamental great circle is the family of secondary great circles, the **MERIDIANS.** They also have the center of the earth as their center, but go both through the N and S poles, and intersect the equator at right angles.

The **GEOGRAPHICAL LATITUDE** ϕ of a location on the earth's surface is the angle between the equator and this location, measured along this location's meridian.

The latitude of the equator $= 0°$
N-pole $= 90°$ N
S-pole $= 90°$ S

All locations with the same latitude lie on a latitude circle or a **PARALLEL,** parallel to the earth's equator, hence the name. The parallels are not great circles, their center is **NOT** the center of the earth. It lies on the axis of the earth, either below or above the center.

The geographical **LONGITUDE** of a location is the angle between a certain point on the earth's equator with longitude 0°, and the point where the location's meridian intersects the equator. The choice of the 0° longitude meridian is arbitrary. For historical reasons, the meridian which goes through the observatory in Greenwich (London), England was chosen to have 0° longitude.

Instead of specifying an angle from 0° all the way around to 360°, Longitude is expressed either from 0° to 180° East of Greenwich or from 0° to 180° West of Greenwich. The 180° meridian is the so-called **INTERNATIONAL DATELINE.** All locations with the same longitude lie, of course, on the same meridian and have the same time, as we shall see later.

So far, so good, but how do we measure our latitude and longitude? We shall assume for the following considerations that the earth is a perfect sphere. We will

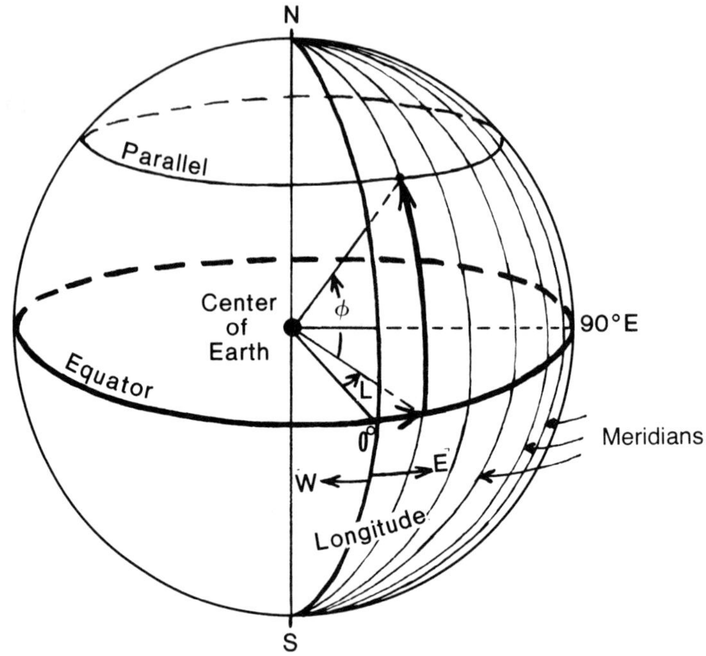

Figure 1

neglect its "oblateness" which causes a slight deviation of geographical and geocentric coordinates (not more than about 12 minutes of arc at latitudes of about 45°, the maximum deviation).

a) **LATITUDE DETERMINATION**

The angle between the horizon and the North star **POLARIS** is your geographical latitude. Here again we make an approximation: Strictly speaking, the angle should be between the horizon and the North Celestial Pole (NCP), Polaris is 0.8° off at present. From now on, we will use **NCP** and **POLARIS** interchangeably. If you are interested where the NCP lies with respect to Polaris, consult the map of circumpolar constellations on page 2–23.

b) **LONGITUDE DETERMINATION**

At the instant of High Noon, (when the sun is highest in the sky, or when the sun's altitude is largest, or when the sun goes through your meridian) you set a clock to 12 hr 00 min noon. It is, in other words, your local (mean) solar time. **IMAGINE, YOU ARE IN GREENWICH, ENGLAND, AT THE OBSERVATORY AND SET YOUR CLOCK.** From now on we will call Greenwich's local (mean) solar time also: **UNIVERSAL TIME (UT),** and to distinguish our clock which shows Universal Time from another one which will show civil or Standard Time, we shall call it our **CHRONOMETER;** as all the captains of ships do whenever they need to find out where they are. Imagine further that we then fly to New Orleans, La. On the next day we observe High Noon in New Orleans and find that it is also 12 noon on our civil clock, showing the same time as all the clocks of all the people in New Orleans. No great surprise! But now look at your chronometer. What does it show? 6 pm. In New Orleans it is High Noon when the sun is setting in Greenwich, we are 6 hr behind. Since the Earth turns 360° in 24 hr, 6 hr correspond therefore to an angle of 90°, or: The longitude of New Orleans is 90° W of Greenwich.

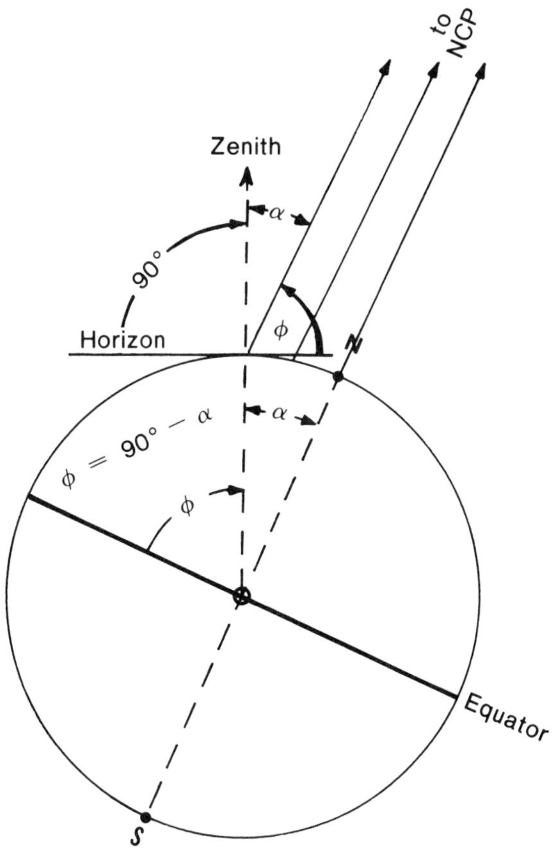

Figure 2

Another example: Suppose you went to Dallas, Texas instead of New Orleans. You may know that both locations lie in the Central Standard Time (CST) zone and you do not change your clock setting if you drive from New Orleans to Dallas.

Suppose it is now High Noon in Dallas. What does our chronometer show? A little after 6 pm, let's say: 6 hr 28 min pm. We are now 6 hr and 28 min behind Greenwich, our longitude is therefore:

6 hr correspond to 90°
4 min correspond to 1°
28 min correspond to 7° or 97° West of G.

What does our civil clock show? Our civil clock goes by Central Standard time: when it is High Noon in New Orleans (at the center meridian of the CST zone), it is not yet High Noon in Dallas. The earth has to turn through 7° more or has to turn for another 28 min until the sun is highest in Dallas. Therefore, at High Noon in Dallas, our civil clock shows 12 hr 28 min pm CST. It is High Noon at 12 hr 00 min on Standard Time clocks only for all those locations which lie on every 15 th meridian; within the continental US:

on the 75 th meridian (EST),
the 90 th meridian (CST)
105 th (MST)
120 th (PST)

For all locations East of these center meridians, High Noon is earlier than 12 noon on the Standard Time clock. For all locations West of these center meridians, High Noon is later than 12 noon on the Standard Time clock.

APPARATUS

Atlas, access to an Earth globe, ruler, protractor with plumb-line, starfinder, clock showing Standard Time.

If specified by your instructor: highway maps, ______________________

____

PROCEDURE

1) Make yourself familiar with the Latitude and Longitude coordinate system. Investigate which countries and which cities lie on the same parallel and on the same meridian. What parts of the earth lie above the arctic circle and the antarctic circle, which parts are touched by the Tropic of Cancer, the Equator, the Tropic of Capricorn, etc. Then answer the questions of part 1) on the question sheet, page 5–31 at the end of this unit. Read the required coordinates from the globe or the atlas with an accuracy of $\pm 1°$. If you use an atlas, include the kind of projection and the scale used for the map you read. If no grid lines go through the listed cities, you must then interpolate. For example, what is the latitude and longitude of location x below? Take a ruler and measure how many cm or inches correspond to the angular distance between grid lines. Do that for both latitude and longitude if necessary; then measure the distances between the grid lines and the location in question. Following this procedure, we find that:

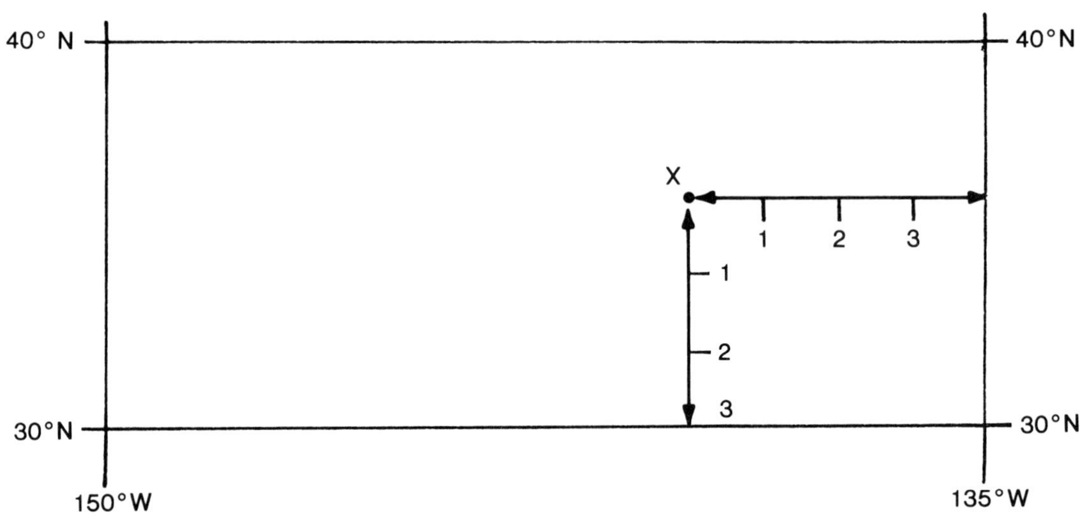

Figure 3

5 cm correspond to a latitude difference (40° − 30°) = 10°
1 cm corresponds to 10/5 = 2°

3 cm therefore correspond to $3 \times \frac{10°}{5} = 6°$. The latitude of x = 36° N

For the longitude: 12 cm correspond to 150° − 135° = 15°

1 cm corresponds to 15°/12

4 cm therefore correspond to $4 \times \frac{15°}{12} = 5°$. The longitude of x = 135° + 5° = 140° W

2) Consult either an atlas, a highway map or any other geographical map with a small enough scale so that you can read the coordinates of your own location accurately. Interpolation, of course, still will generally be required. Answer the questions of part 2) on the question sheet, on page 5–33 at the end of this unit. Include requested information about the map you used for these readings.
3) Consult Fig. 4, the map of the continental USA (excluding Alaska) on p. 5–23 and at the end of this unit. Do the following things in order:
   a) Write in the map the name of the state in which you live and the name of its capital, in the appropriate place.
   b) Mark by a dot your location within the state if different from the capital.
   c) Designate the center meridians of each time zone (75°, 90°, etc) by **RED** lines.
   d) Designate the theoretical borders of each time zone (7.5° either side of the center meridian) by **BLUE** lines.
   e) Draw the **ACTUAL** borders of the time zone you live in. Use pencil or another color. Shade this zone or make it otherwise especially noticeable, and label your time zone. You may consult highway maps or atlases containing this information. It will be very informative and you may want to inquire why the time zone lines were determined in this apparently senseless way.
   f) Draw with different colors all the other time zones within this map and label them.

   g) ______________________________

   ____

   h) ______________________________

   ____

4) a) For the following questions consult the map, Fig. 4 on p. 5–23. Read the required information from it. Answer part 4) on the question sheet, p. 5–34.
   b) There are two states with rectangular outline. List their borders by specifying their latitudes and longitudes.
   c) Answer the questions about differences in latitude and longitude and their corresponding distances on p. 5–34.

Unit 5.4 Name ______________________

Date due ____________ Instructor ______________________ Section ____________ Seat _____

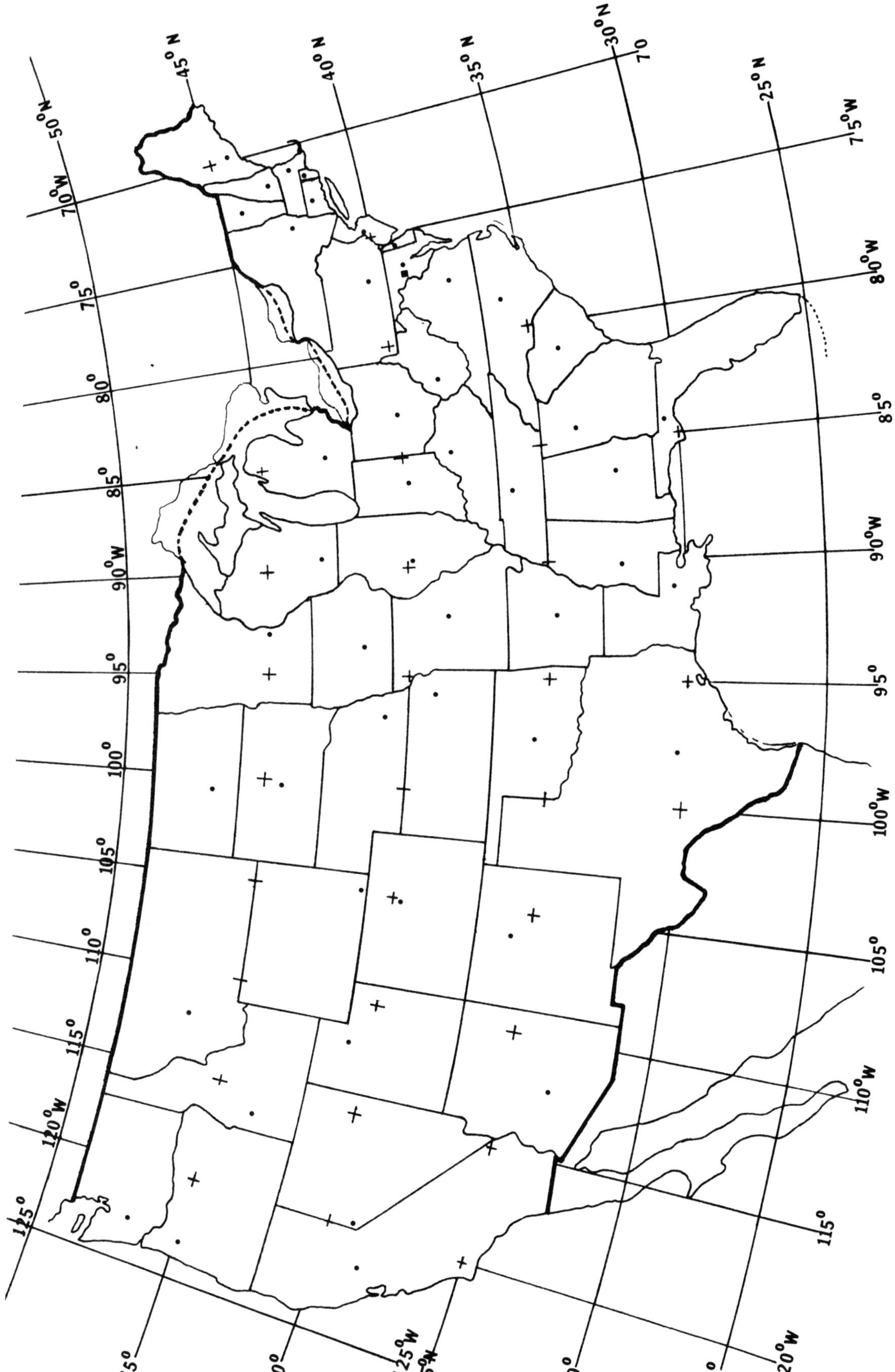

Figure 4

5) During a year (the time the Earth needs to make one complete revolution around the Sun) the axis of the Earth points in the same direction; its equator is inclined by 23.5° to the ecliptic plane, the obliquity of the ecliptic. This inclination causes the seasons. Note the incidence of sunlight on the Earth particularly in spring or fall. On the equinoxes (March 21 or September 22) the Sun appears in the Zenith on all points of the Earth's equator at High Noon, while the Earth rotates around its axis. In the drawing below the Sun rays would be perpendicularly incident on the plane of the paper. Imagine, your eye is the Sun!

The two drawings, Fig. 7 and Fig. 8, p. 5–26 and 5–27 **DO NOT SHOW** this inclination of the Earth's equator to the ecliptic. Note, however, that the Sun is in the zenith at High Noon at a certain location on the Earth's equator. The date is one of the equinoxes. Study these two drawings! Fig. 7 is the Earth as seen from the top, Fig. 8 is a side view of the Earth. Both views are from out in space, so to speak. Some major cities are correctly plotted, both in Latitude and Longitude. For the date of the year the drawings Fig. 7 and 8 show, the Sun is at High Noon in the zenith on the equator at location 0° longitude, just below Accra, Ghana. The Sun goes through the meridian at all points with 0° longitude, it is therefore 12 noon in Greenwich, as on all points of this meridian. The Greenwich (local, mean) solar time is: 12 noon—and so is the Universal Time. Our chronometer shows 12 noon UT wherever we are on the Earth, as drawn.

Look at the **TOP-VIEW** map and note the arrow around the N-pole, indicating the Earth's rotation. Now consider New Orleans, La.: At the same time the Sun goes through Greenwich's meridian, the Sun is just rising above the horizon in New Orleans. It is 6 am as read on our civil clock, 12 noon on our chronometer, i.e. New Orleans' longitude is 90° W. In Dallas, Texas, the Sun has not yet risen. It is also 6 am on our civil clock since both New Orleans and Dallas lie in the Central Standard Time zone. We will have to wait 28 min until the Sun will rise. If we wait, then: at 6 hr 28 min am on our civil clock

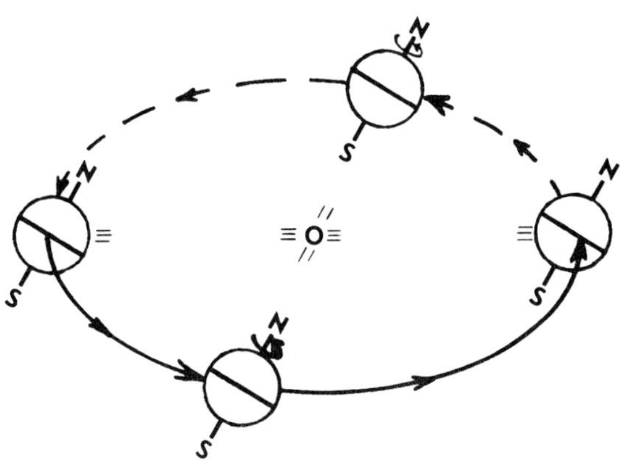

Figure 5

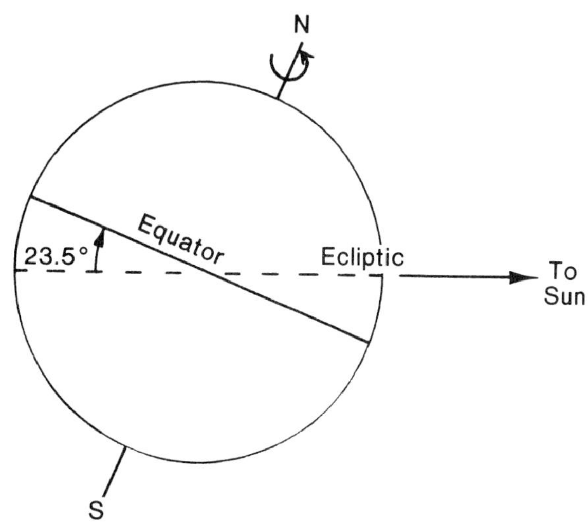

Figure 6

Abbreviations used for City names:

(from 0° longitude westward)

G	Greenwich, England
A	Accra, Ghana
Dak	Dakar, Senegal
NY	New York, N.Y.
M	Miami, F.
Ch	Chicago, Ill.
NO	New Orleans, La.
D	Dallas, Tx.
MC	Mexico City, Mexico
SF	San Francisco, Cal.
F	Fairbanks, Alaska
H	Honolulu, Hawaii
T	Tokyo, Japan
Ma	Manila, Philippines
UB	Ulan Bator, Outer Mongolia
Dac	Dacca, Bangladesh
Mo	Moscow, USSR
Na	Nairobi, Kenya

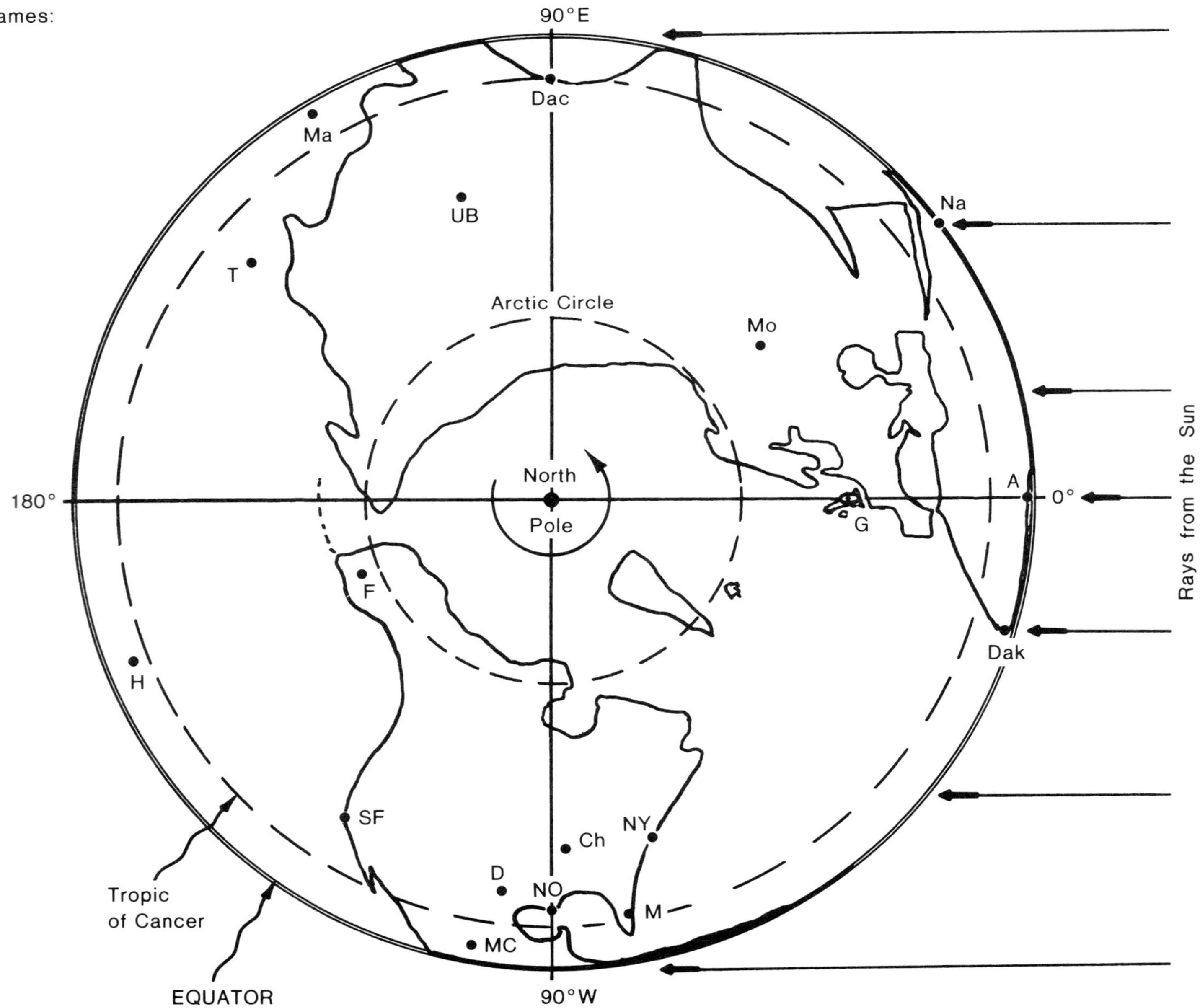

Figure 7

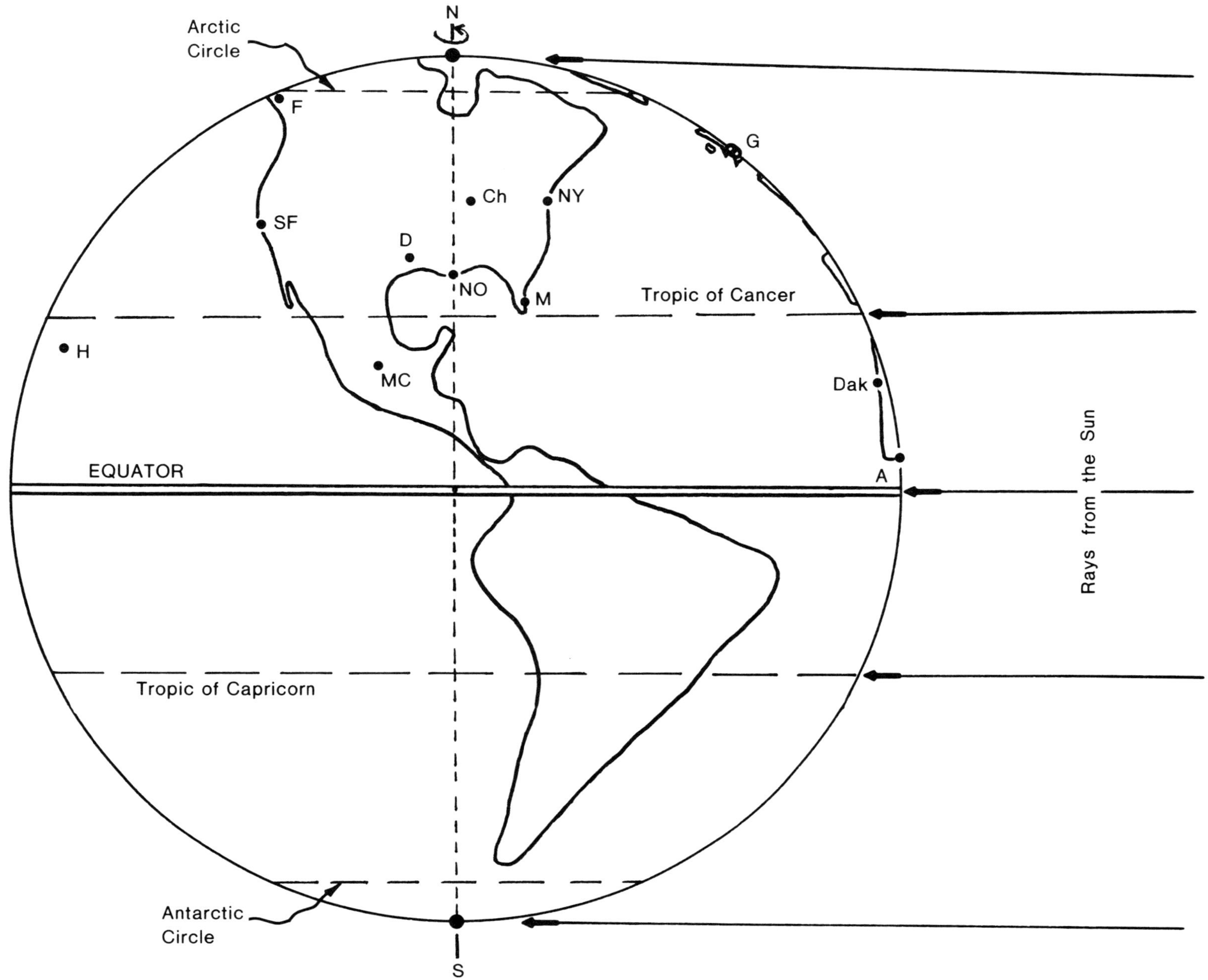
N
Arctic Circle
F
SF
D
Ch
NY
NO
M
G
Tropic of Cancer
H
MC
Dak
A
EQUATOR
Rays from the Sun
Tropic of Capricorn
Antarctic Circle
S

Figure 8

the Sun will rise in Dallas, our chronometer will read: 12 hr 28 min pm UT, meaning the Sun is 28 min past High Noon in Greenwich. The longitude of Dallas is therefore:

$$\begin{array}{r} 6 \text{ hr} = 90^\circ \\ 28 \text{ min} = \ \ 7^\circ \\ \hline 6 \text{ hr } 28 \text{ min} = 97^\circ \text{ W of Greenwich} \end{array}$$

On the question sheet, p. 5–35, fill the empty spaces in part 5). Note that for some locations the UT is given, for some others their longitude and latitude. You are welcome to draw in the **TOP VIEW** map the universal times, time zones, etc. Another, identical set of maps is enclosed at the end of this unit for your personal use. You may use Fig. 8 (Side View) also to determine the altitude of the sun at High Noon. A sample construction and questions are on the question sheet, p. 5–36.

6) **MEASUREMENT OF YOUR GEOGRAPHICAL LATITUDE.**
   a) If you use the starfinder, commercially or self-made, align the RA circle horizontally, center Polaris in the viewing tube, and read the angle on the Decl. scale. This angle is the angle between the horizon and Polaris, your geographical latitude.
   b) If you use the protractor of Unit 3.1, attach a thin string through the center of the protractor and something heavy at the end of the string. If you look along the edge of the protractor to Polaris (or through an attached straw), the angle between Polaris and the plumb-line is: 90° + your latitude, see diagram below. Answer the questions of part 6) on the question sheet, p. 5–37.

7) **MEASUREMENT OF YOUR LONGITUDE.**

   If you had a UT chronometer available, and could determine when the Sun goes through your meridian, you would then read UT time and the conversion to longitude would be done as explained on p. 5–19.

   Assume you know nothing about your longitude except that your own clock shows Standard Time, that you do know in which time zone you live and that the longitude of the center meridian of EST is 75° W, of CST is 90° W, of MST is 105° W, and PST is 120° West.

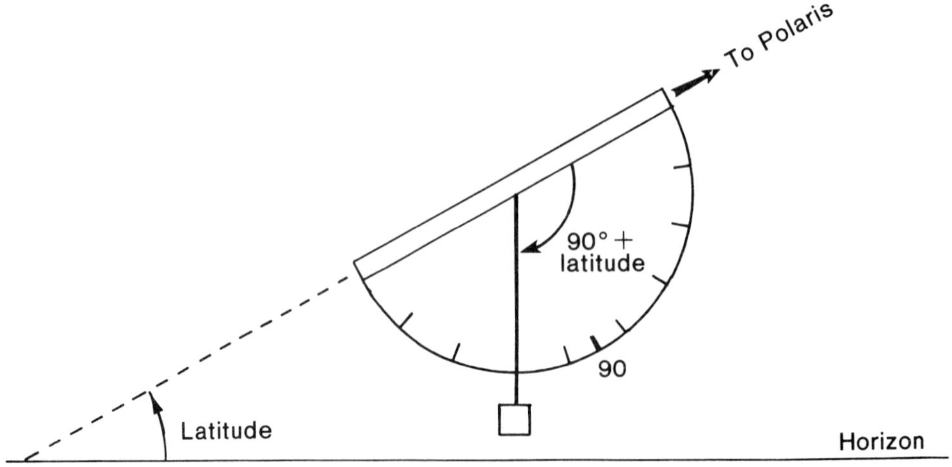

Figure 9

a) You also know the N-S direction. Note the instant the Sun goes through your meridian (N-S). You may draw the N-S direction with chalk e.g. on the ground, have a stick perpendicular to the ground at the S end of the N-S line and observe when the stick's shadow crosses the N-S line. As you already know, if the Sun crosses your meridian at 12 hr 00 min noon, you live on a time zone's center meridian. In general, you will observe that High Noon is either before or after your Standard Time clock shows 12 hr 00 min noon. Your longitude is therefore the number of degrees E or W of the center meridian, corresponding to the time difference between your High Noon and 12:00 on the Standard Time clock. Answer part 7) on the question sheet, p. 5–39.

b) Assuming you do **NOT** know the N-S direction.
   1) When the Sun is at High Noon = going through your meridian, then the Sun is as high above the horizon as it will ever be on that day, the shadow of a stick planted perpendicularly in the ground is therefore shortest. It is difficult to judge this instant of shortest shadow, unless you keep track of the slowly decreasing and then slowly increasing length of the shadow for some time before and after High Noon and note the time on your clock as indicated in the drawing below. With a little bit of thought and analysis you will come to the correct conclusion as to the instant of High Noon.
   2) You can always find the direction to Polaris at night. The line from the N-point directly below Polaris to your position is then a part of your meridian, or the N-S line.

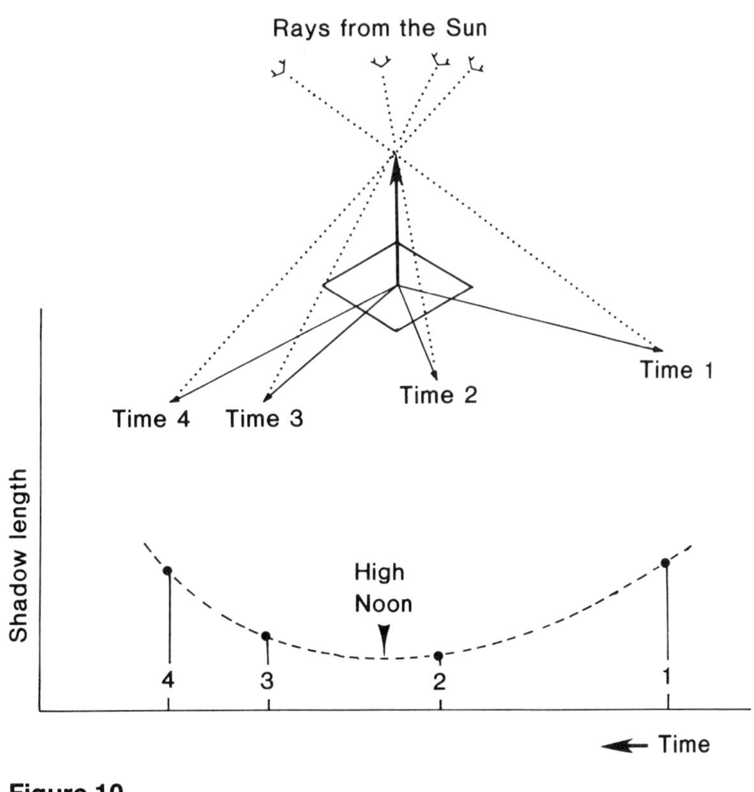

Figure 10

3) You may know the position of the Sun in celestial coordinates (Decl. and RA) on the day you do this measurement. Then, at night you can watch for the instant, a star with RA = sun's RA + 180° goes through your meridian. High Noon then would be 12 hr earlier or later. Remember, that we use our civil clock for these measurements. Civil time is based on solar time, and you may know that sidereal (stellar) time is different. So, what changes if we use a star for our longitude measurement? If we kept a clock with sidereal time, we would note that a sidereal day is about 4 minutes shorter than a solar day. Does this argument apply here? On the other hand, the sun moves through the ecliptic (once through all the Zodiac belt constellations in 1 year or 365 days. How much does the sun move during 1 day? about 1° relative to the stars. How much of a mistake do we make if we measure our longitude by this procedure 3)?

Unit 5.4 Name ____________________

Date due ____________ Instructor ____________________ Section ____________ Seat ______

Part 1

Atlas used: ____________________

Map scale and projection: ____________________

List all countries, from N to S, as specified:

1) The 0° meridian goes through them:

____

____

2) The ____ W meridian goes through them:

____

____

3) The ____ E meridian goes through them:

____

____

List all countries through which the specified latitude circles (parallels) go, from 0° longitude eastward:

4) 60° N: ____________________

5) Tropic of Cancer: ____________________

____

6) Equator: ____________________

____

7) Tropic of Capricorn: ____________________

____

8) Arctic Circle: ____________________

____

9) ____________________

____

10) ____________________

____

Give the coordinates, Latitude and Longitude, with an accuracy of ±2° of the following cities:

City	Lat.	Long.	City	Lat.	Long.
11) Washington, D.C.			19) Rio de Janeiro, Brazil		
12) Juneau, Alaska			20) Murmansk, USSR		
13) Oslo, Norway			21)		
14) Moscow, USSR			22)		
15) Cairo, Egypt			23)		
16) New Delhi, India			24)		
17) Canberra, Australia					
18) Little America, Antarctica					

In which of the cities, listed above, **WILL THE SUN**

25) never be in the zenith: ______________________________

____

26) not set between March 21 and September 22: ______________________________

27) not set between Sept. 22 and March 21: ______________________________

28) be in the zenith at High Noon at least once in a year: ______________________________

____

29) ______________________________

____

____

30) ______________________________

____

____

____

Unit 5.4 Name ____________________

Date due __________ Instructor ____________________ Section __________ Seat _____

Part 2

Read from a small scale map the Latitude and the Longitude of the following locations as accurately as possible. Map used: ____________________

Location	Latitude	Longitude
your school		
2) intersection of ____________________ with ____________________		
3) intersection of ____________________ with ____________________		
4) ____________________		
5) ____________________		

6) The distance between location 2) and 3) (as read from the map) is: __________ [miles].

7) The difference in longitude between location 2) and 3) is: ______ [°]
Their difference in latitude is either 0° or negligibly small (for our purposes).

8) Divide the distance in miles from 6) by the difference in longitude in degrees from 7)

$$\_\_\_\_\_\_ \left[\frac{\text{mi}}{\text{degr}}\right] = \_\_\_\_\_\_$$ [miles/degree] at latitude: __________ ° N

9) Suppose two locations, separated by the same difference in longitude, as 2) and 3) above were on the earth's equator. Would they also be separated by the same distance in miles as are 2) and 3) at the above latitude?

mark ( ) YES ( ) NO

10) Suppose, one bear stands at longitude 90° E and another bear at longitude 90° W, but the actual distance between them is only 10 feet, what is:

a) their color: ____________________

b) their latitude: almost exactly __________ ° _____

11) The equatorial circumference of the earth is 25,000 [mi] or 40,000 [km]. The longitude of the equator ranges from 0° to 360°.

Therefore, 1° longitude difference on the equator corresponds to ______ [mi] = ______ [km].

Unit 5.4 Name ____________________

Date due ____________ Instructor ____________________ Section ____________ Seat ______

Part 4

1) Name the four states which meet each other in one point:

a) ____________ b) ____________ c) ____________ d) ____________

The coordinates of this point are:

geogr. latitude: ____________ , geogr. longitude: ____________ .

2) Name the two states, rectangular in outline, from N to S:

a) ____________________ b) ____________________

Fill in the empty spaces in the table below, using state a) and b) in the same sequence as above:

State	Latitude of northern border (line from east to west)	Latitude of southern border (line from east to west)	Longitude of eastern border (line from north to south)	Longitude of western border (line from north to south)
a)				
b)				

	Latitude	Longitude
distance A, in mi between N and S border of state b)		
difference B in degrees between N and S borders of state b)		
$\frac{\text{distance A}}{\text{difference B}}$ = mi per degr. Lat.		
distance C, in mi between E and W border of state b)		measured along northern border: ____________ southern border: ____________
difference D in degrees between E and W border of state b)		
$\frac{\text{distance C}}{\text{difference D}}$ = mi per degree Long.		for north. border: ____________ for south. border: ____________

Unit 5.4 Name ____________________

Date due __________ Instructor ____________________ Section __________ Seat _____

Part 5

Study Fig. 7 and Fig. 8.

At the equinoxes the sun is in the zenith at a point on the equator (Lat. 0°) with Longitude 0°, just south of Accra, Ghana as shown in both Fig. 7 and Fig. 8. For all locations on the 0° meridian it is local High Noon, 12 hr 00 min noon. Since the 0° meridian also goes through Greenwich, England, it is also 12 hr 00 min noon Universal Time (UT) for the instant the drawings show, not only on the 0° meridian but everywhere on earth.

Fill in the empty spaces in the table below, read the latitude from Fig. 8 with a protractor (you may have to draw a parallel through the location in question), calculate either the longitude if the time is given, or the times if the longitude is given, always for the instant shown in the figures.

Location	local solar time hr	min	a or pm	Standard Time hr a or pm	UT am or pm	Latitude in degrees	Longitude in degrees
1) Greenwich, England	12	00	noon	12 noon	12 noon	52°N	0°
2) Accra, Ghana					12 noon	4°N	0°
3) Dakar, W. Africa	10	50	am			14.5°N	
4) New York, N.Y.							74°W
5) New Orleans, La.	6	00	am	6 am CST		30° N	
6) Chicago, Ill.							87°42'W
7) Dallas, Tx.							97°W
8) San Francisco, Cal.	3	50	am				
9) Fairbanks, Alaska							147°W
10) Honolulu, Hawaii	1	28	am			20°N	
11) Tokio, Japan							140°E
12) Dacca, Bangladesh	6	00	pm		12 noon		
13) Moscow, USSR	2	29	52 sec pm				
14)							
15)							
16)							
17)							
18)							
19)							
20)							

21) If you would like to complete the map of South America in Fig. 8, read the latitudes and longitudes of some major cities and plot them. To do this is not as easy as it reads. You can easily plot the location's latitude circle in the Side View map. You can also easily plot the location's meridian in the Top View map, imagining the South Pole at the N-pole's location. The actual location in either map must somehow be the intersection of a location's latitude circle with its meridian. Try it for the following cities, using a colored pen or pencil. You may find that the outline of the South-American continent is rather crudely drawn and you may find Rio to be in the ocean.

a) Caracas, Venezuela, b) Recife, Brazil c) Rio de Janeiro, Brazil
d) La Paz, Bolivia e) Buenos Aires, Argentina f) Punta Arenas, Chile

22) How high above the horizon is the Sun at High Noon (its altitude) on a certain day of the year at a certain location with latitude ϕ?

The question is answered below for the summer Solstice, June 22, for a location with latitude 50° N.

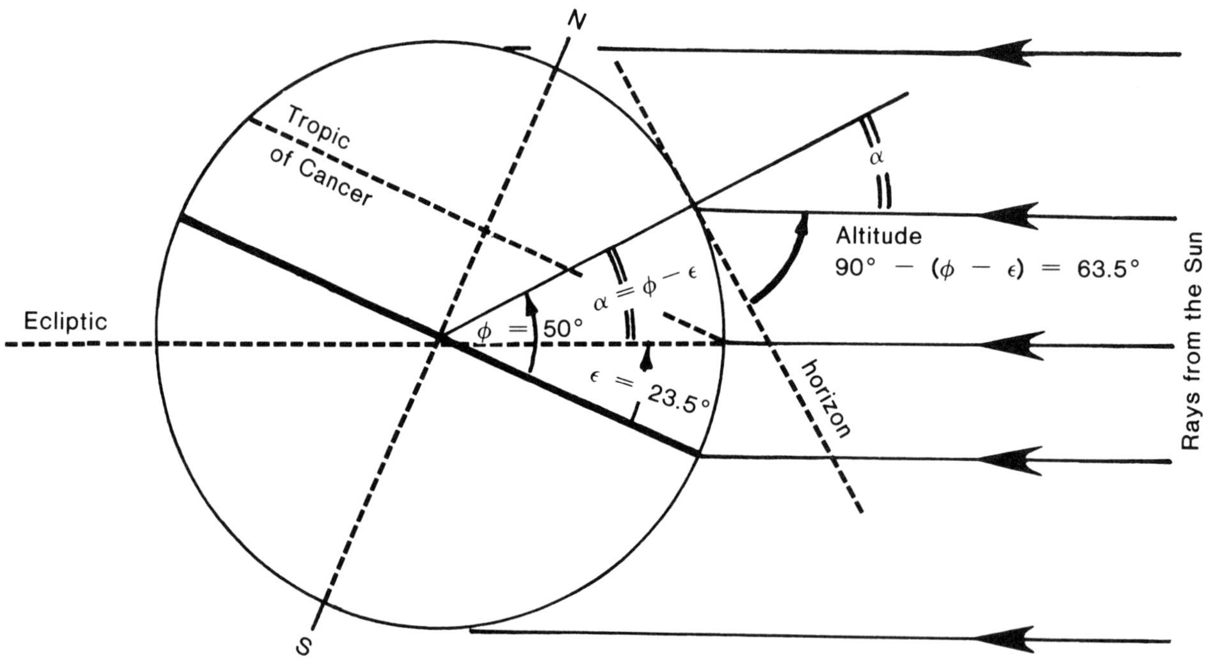

Figure 11

Fill in the empty spaces in the table below for the cities and dates to be specified:

	Altitude of the Sun at High Noon (Local 12 hr 00 min Noon) on the Dates:		
Location	March 21 or Sept. 22 vern. or aut. equinox	June 22 summer solstice	December 22 winter solstice
1)			
2)			
3)			

Unit 5.4 Name ____________________

Date due ____________ Instructor ____________________ Section ____________ Seat ______

Part 6

MEASUREMENT OF YOUR GEOGRAPHICAL LATITUDE

1) The angle between the **HORIZON AND POLARIS** was measured with what instrument:

____

2) The smallest angle which can be accurately read on your instrument is ? Please mark:
a) 0.1° b) 0.5° c) 1° d) 2° e) more than 2°

3) Measure the angle between horizon and Polaris 4 times in a row, write the values in the table below, and then take the average.

measurement	angle
1	
2	
3	
4	

sum of the 4 measurements: ____________ sum/4 = ____________

4) Your geographical latitude is: [ ]

5) Suppose you had used an instrument which allowed you to read an angle with an accuracy of 0.1°, and suppose further that you had measured the angle between **POLARIS AND THE HORIZON** with this accuracy. What would be the maximum error in your latitude angle stated above if you had measured the angle between the **NCP AND THE HORIZON?**

____

6) If you do use Polaris for the angle measurement, you can still prevent this error from occurring. Describe, and remember that Polaris is a circumpolar star:

____

____

____

Make a sketch, if you like:

Unit 5.4 Name ______________________________

Date due ____________ Instructor ____________________ Section __________ Seat ____

Part 7

MEASUREMENT OF YOUR GEOGRAPHICAL LONGITUDE

In order to measure your longitude with a fair amount of accuracy, you must know: a) how accurate is your N-S direction, and b) how accurate is your clock (showing Standard Time).

1) Describe how you determined your North-South direction (part of your meridian):

____

____

____

____

2) How accurate is your clock, and by what standard did you set it. Describe:

____

____

____

3) The sun went through your celestial meridian (its shadow passed through your meridian) when your standard time clock showed: ____________________

4) Your standard time zone is: __________ The center meridian of this time zone is at longitude: ____

5) High Noon at your location is therefore __________ min a) earlier b) later MARK! than High Noon on the center meridian of your time zone.

6) Your time zone is __________ hr West (later) of Greenwich.

7) Taking 5) and 6) together, your longitude is: [ ]

8) Comment on: ______________________________

____

____

____

____

Unit 5.4 Name ____________________

Date due __________ Instructor ____________________ Section __________ Seat _____

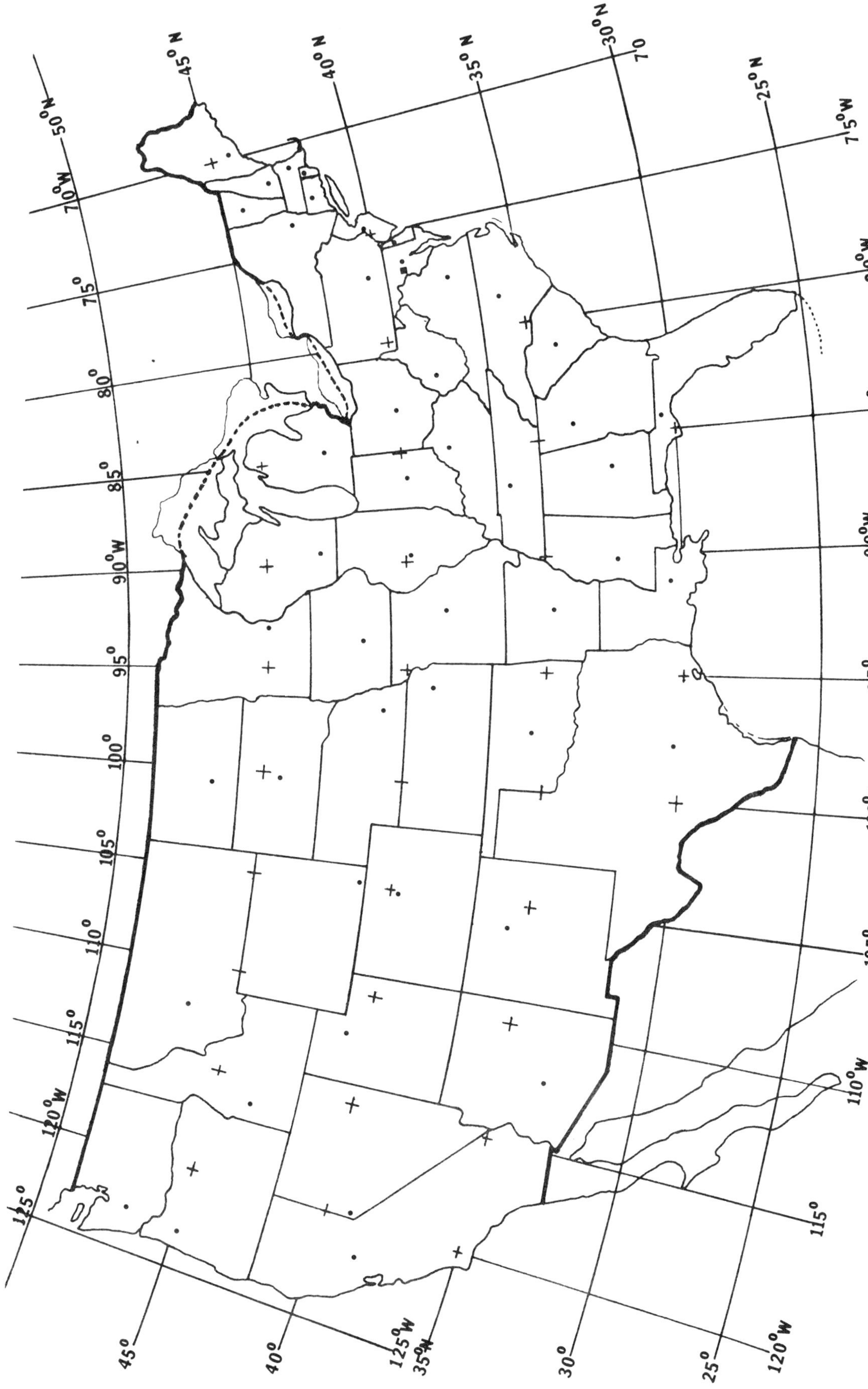

Figure 4

Abbreviations used for City names:

(from 0° longitude westward)

G	Greenwich, England
A	Accra, Ghana
Dak	Dakar, Senegal
NY	New York, N.Y.
M	Miami, F.
Ch	Chicago, Ill.
NO	New Orleans, La.
D	Dallas, Tx.
MC	Mexico City, Mexico
SF	San Francisco, Cal.
F	Fairbanks, Alaska
H	Honolulu, Hawaii
T	Tokyo, Japan
Ma	Manila, Philippines
UB	Ulan Bator, Outer Mongolia
Dac	Dacca, Bangladesh
Mo	Moscow, USSR
Na	Nairobi, Kenya

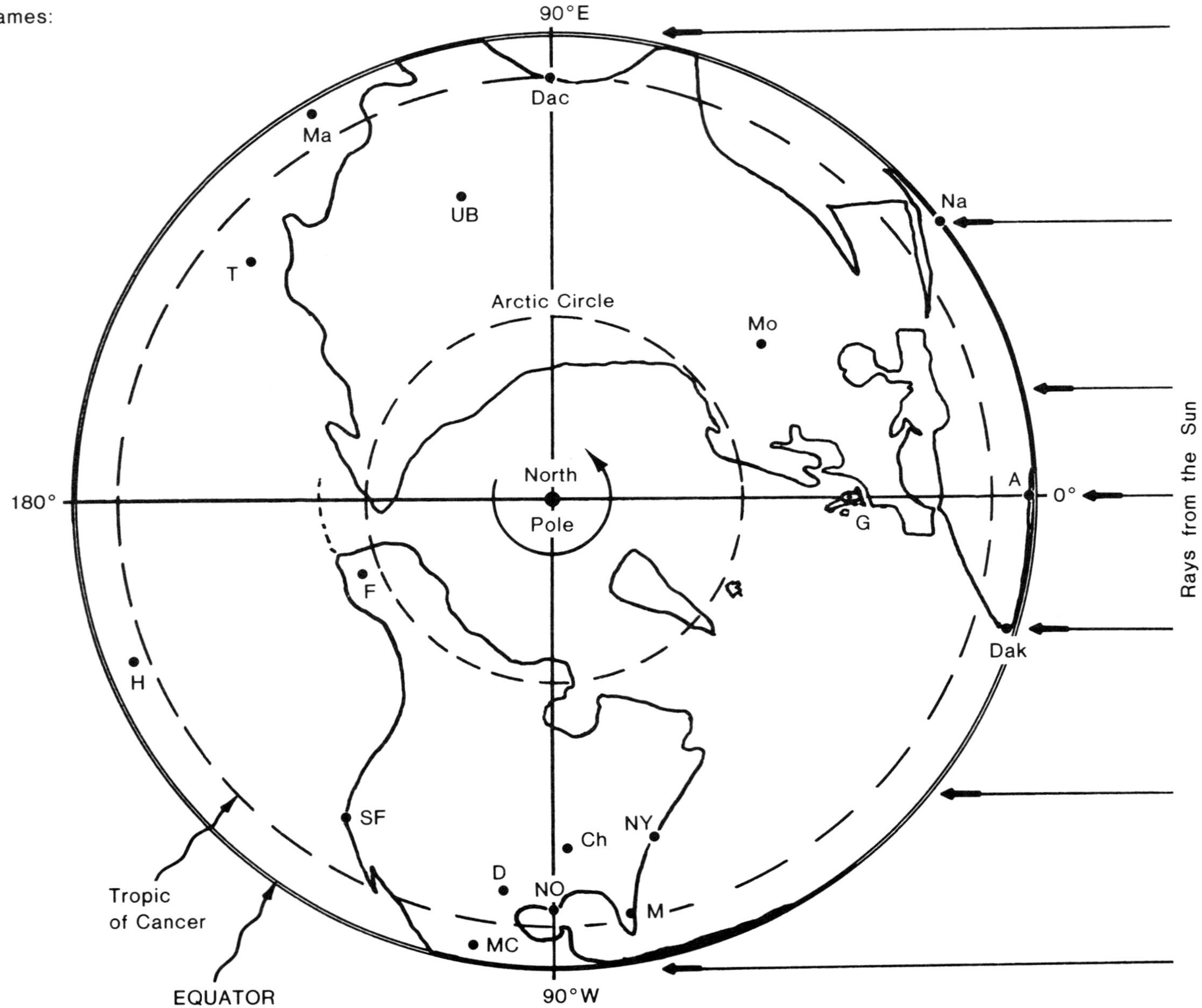

Figure 7

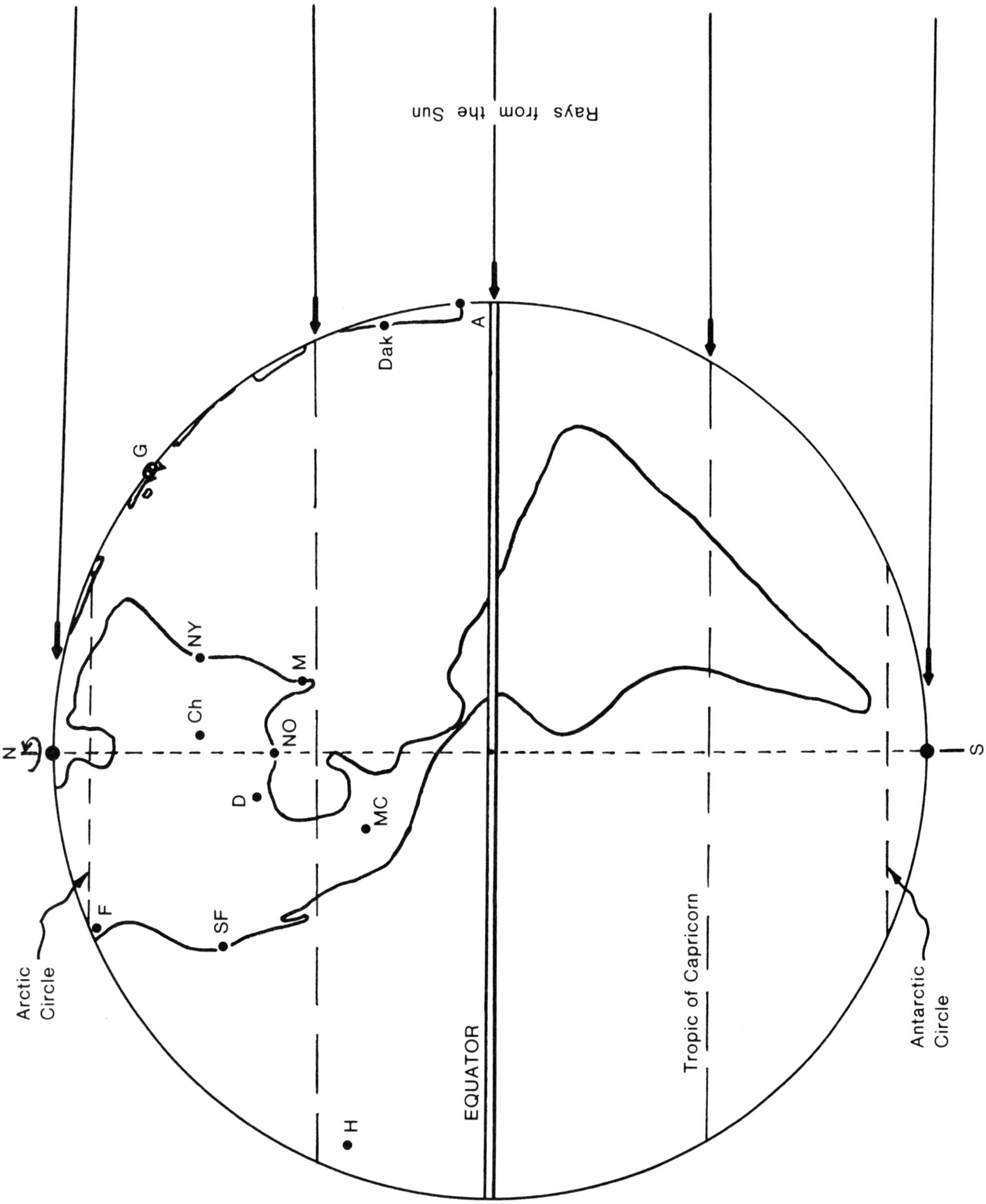
Rays from the Sun
A
Dak
G
NY
M
Ch
N
NO
S
D
MC
F
SF
H
Arctic Circle
EQUATOR
Tropic of Capricorn
Antarctic Circle

Figure 8

Unit 5.5

The Celestial Sphere: Right Ascension and Declination

OBJECTIVE

To become familiar with the "equatorial coordinate system" and to use Right Ascension and Declination to locate celestial objects.

INTRODUCTION

The Right Ascension (RA) and Declination (Dec) system is the astronomical equivalent to the geographical Longitude and Latitude system. The Longitude is measured along the Earth's equator from an arbitrary point, the Greenwich meridian. The RA is measured along the **CELESTIAL** equator from an arbitrary point, the **VERNAL EQUINOX** (VE). Earth Latitude is measured as the angle from Earth's equator to the location in question, along the location's meridian. Declination is measured as the angle from the **CELESTIAL** equator to the location in question, along the location's **HOUR CIRCLE.**

The fundamental great circle of the RA and Dec system is the celestial equator with its pole, the North Celestial Pole (NCP). The center of the circle is the center of the Earth, the plane of the circle is perpendicular to the axis of the Earth. While the Earth's axis goes through the N- and S-pole of the Earth, its extension goes through the NCP and the SCP (South Celestial Pole). At right angles to this fundamental great circle is the family of secondary great circles, the **HOUR CIRCLES.** The special hour circle which passes through the observer's zenith is called the **CELESTIAL MERIDIAN**, the extension into space and onto the celestial sphere of the observer's (terrestrial) meridian. The center of the Earth is also the center of the hour circles. As all meridians pass through Earth's N- and S-poles, so the hour circles pass through the NCP and the SCP.

While the choice of Earth's 0° meridian was arbitrary from a scientific point of view, the choice of 0° or 0 hr RA is not wholly arbitrary:

The Earth spins around its axis which is fixed in direction relative to the distant stellar background (neglecting precession). The celestial equator, always at right angles to the axis of the Earth, is therefore also fixed in space. At the same time the Earth revolves around the Sun in a plane which is inclined to the equatorial plane. Therefore, there are two points where the equatorial plane intersects the plane of the ecliptic. These two points are the EQUINOXES. One of them, the vernal equinox, is the origin of the RA scale.

The angle of inclination between the equatorial plane and the ecliptic plane is the so-called OBLIQUITY OF THE ECLIPTIC. Its value is 23.5°. Another, equally valid, definition of the ecliptic is: The ecliptic is the apparent path (as seen from the Earth) of the Sun relative to the stellar background. It is centered on the belt of the 12 Zodiac constellations.

The relationships between all the quantities described above is shown in Fig. 1. All stars and other celestial objects appear to be located on the CELESTIAL SHPERE. Its radius is extremely large compared to the distances within the solar system, let alone distances on Earth. In Fig. 1 you should imagine the Earth to be just a point at the center of the celestial sphere. The Earth is drawn in Fig. 1 relatively large only for the reason that you may see the relationship between Earth's equator and the celestial equator as well as the relationship between N-pole and NCP.

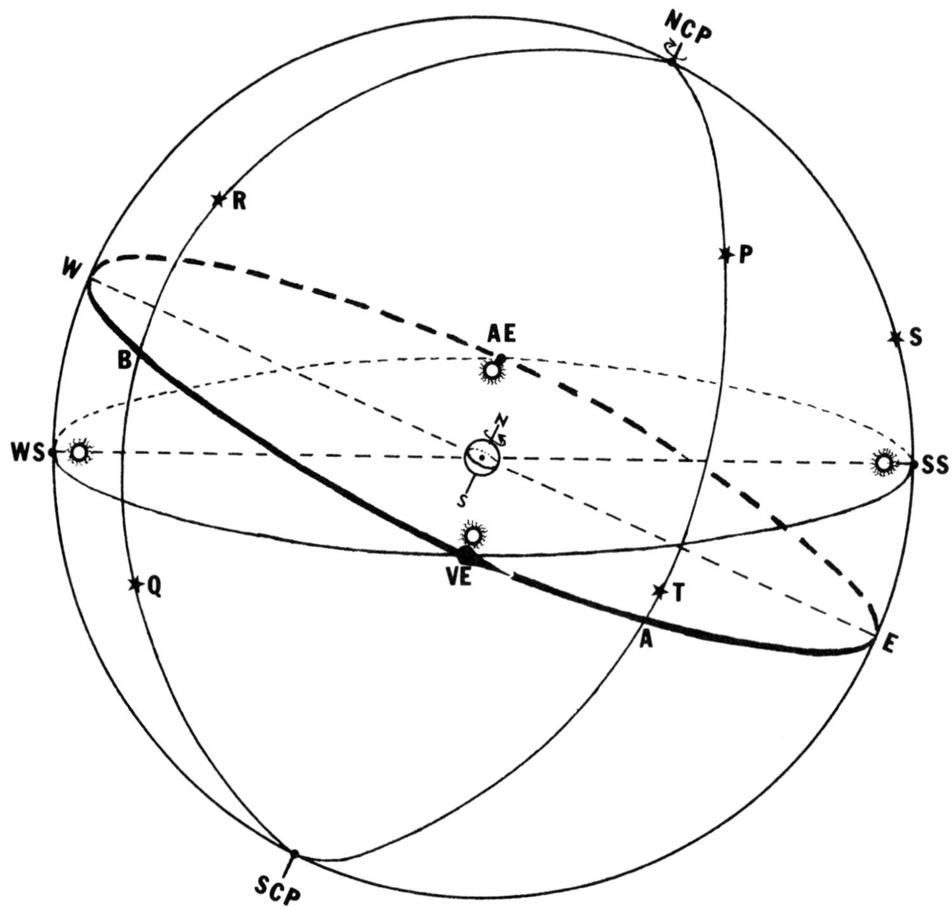

Figure 1. The Equatorial Coordinate System

The following abbreviations are used in Fig. 1

NCP	North Celestial Pole	VE	Vernal (or spring) Equinox
SCP	South Celestial Pole	AE	Autumnal Equinox
SS	Summer Solstice	E	East
WS	Winter Solstice	W	West

The Earth rotates as indicated by the small arrow at the Earth's N-pole from West to East. The celestial sphere seems to rotate in the opposite direction from East to West, as indicated by the small arrow on top of the NCP. (You know that stars rise somewhere on the eastern horizon and set somewhere on the western horizon.) The position of the Sun is shown for both equinoxes and solstices on the ecliptic, in front of points VE, SS, AE, and WS.

Study Fig. 1: Note the Earth's N and S poles and the NCP and SCP at the intersection of the extension of Earth's axis with the celestial sphere. The fat ellipse is the celestial equator, the extension of Earth's equatorial plane. At an angle of 23.5° is the ecliptic. (Measure the angle with a protractor!) The two points of intersection between the equatorial and the ecliptic plane are the equinoxes, with the vernal equinox in front and the autumnal equinox below the plane of the paper. The RA of an object is measured **ALONG THE CELESTIAL EQUATOR, STARTING WITH** 0° or 0 hr **AT THE VE IN AN EASTERLY DIRECTION.** It is expressed either in degrees, from 0° to 360°, or in hours (sidereal) from 0 hr to 24 hr.

The Dec of the celestial equator or of any object on the celestial equator is 0°. The Dec of the NCP is +90°, that of the SCP −90°. + or − are used instead of N and S.

EXAMPLES

1) Consider the Sun to be at point WS on the ecliptic. Earth's southern hemisphere would receive more light than the northern one, the S pole is in sunlight, the N pole in shade. It is winter on the northern hemisphere, the Sun is in the Winter Solstice (WS), the date would be December 21. What are the Sun's coordinates?

   Answer: The Sun's hour circle is NCP-W-WS-SCP. It intersects the celestial equator at point W. The RA of the Sun is then the arc VE-E-AE-W corresponding to 270° or 18 hr. The Dec of the Sun is the arc W-WS or −23.5°. (− because it is located below the celestial equator).

   During the next quarter year the Earth revolves one quarter the circumference around the Sun or the Sun seems to move one quarter around the ecliptic from point WS to VE. It is then March 21, the vernal equinox. The coordinates of the VE are: RA = 0° or 0 hr; Dec = 0°, because the VE is on the celestial equator.
2) What are the coordinates of star P? Answer: Star P's hour circle is NCP-P-A-SCP. It intersects the celestial equator at point A. The RA of star P is therefore the arc VE-A, or about 30° or 2 hr as judged from the drawing. The Dec of star P is the arc A-P (along P's hour circle). As judged from the drawing, it is about 50°. (+ because it is located above the celestial equator).

EXERCISES

1) Express the location of the objects listed in RA and Dec, on p. 5–49.
2) Plot the Messier objects listed below by RA and Dec on the appropriate star maps on pages 9–37 to 9–45 in the appendix.

   Use the following symbols for the plot:

⁘	open or galactic cluster	⁙	globular cluster
	diffuse emission nebula		diffuse absorption nebula
○	planetary nebula		elliptical galaxy
	spiral galaxy		

   Plot the Messier objects: ______________________________

   ____

   ____

Unit 5.5 Name ______________________________

Date due ______________ Instructor ______________________ Section ____________ Seat _____

Read from Fig. 1, p. 5–46 the coordinates, RA and Dec, express them in arcs, e.g. VE-A, and give an approximate value of the angles in hours and/or degrees.

Object	Right Ascension			Declination	
	arc	estimated value in degrees	estimated value in hr	arc	estimated value in degrees
1) autumnal equinox AE					
2) summer solstice SS					
3) star Q					
4) star R					
5) star S					
6) star T					
7) south celestial pole SCP					
8) point E					
9)					
10)					

Unit 5.6

The Rotation of the Celestial Sphere

OBJECTIVE

To use a starfinder to measure the daily rotation of the celestial sphere.

INTRODUCTION

Because of the rotation of the Earth about its axis, the sky appears as a rotating sphere with the stars fixed to this celestial sphere. The apparent rate of rotation of these objects allows one to measure the rate of rotation of the Earth on its axis.

APPARATUS

A starfinder.

PROCEDURE

1) Align your starfinder with Polaris as described in the first steps of Unit. 3.3. However do not set the R.A. scale with a reference star.
2) Choose a bright object in the Southern sky and center it in the viewing tube. Rotate the R.A. scale until the reading is zero, and record the time for the zero reading.
3) The following measurements will require two to two and a half hours. During this time you are not to change the Declination setting or rotate the R.A. scale. After 30 minutes rotate the viewing tube to recenter the object in the viewing tube. Record the reading from the R.A. scale, and the time of the measurement to the nearest degree and minute of time.
4) Repeat step 3 every half hour for a period of at least two hours.
5) From your data table, calculate the angle moved for each interval, and the total angle moved.
6) Determine the total elapsed time.
7) Calculate the rate by dividing the total angle moved by the total elapsed time. (Recall that to divide by time you must change hours-minutes into hours).
8) Calculate the rate per minute, i.e. divide by 60.

Unit 5.6 Name ____________________

Date due __________ Instructor ____________________ Section __________ Seat ______

Time	R.A. Scale Reading	Angle Moved

1) Total angle moved ____________________

2) Total elapsed time ____________________

3) Rate of movement ____________________ degrees/hour

____________________ degrees/minute

Unit 5.7

The Declination and Right Ascension of Stars and Planets

OBJECTIVE

To allow the student to practice measuring the Right Ascension and Declination of some known objects whose coordinates are easy to verify and hence give the student practice with the starfinder or cross-staff.

INTRODUCTION

Every celestial object has two coordinates that will locate the object on the celestial sphere. The two coordinates used in this exercise are R.A. and Dec. discussed previously, see Unit 5.5. If the coordinates of an object are known, the object can be plotted on the star maps or conversely, by consulting star maps or other reference material, the object can be identified.

APPARATUS

Starfinder or Cross-staff.

PROCEDURE

1) Set up your starfinder to measure Right Ascension and Declination as described in Unit 3.3.
2) After you are sure that your starfinder is reading correctly measure the R.A. and Dec. of ____________ of the objects listed in the data table.

Unit 5.7 Name ____________________

Date due ________ Instructor ____________ Section ________ Seat ____

DATA

Date ____________ Time ____________

Object	R.A.	Dec.
Alpha Andromedae		
Beta Cassiopeiae		
Rigel		
Procyon		
Castor		
Pollux		
Capella		
Arcturus		
Mizar/Alcor		
Vega		
Deneb		
Fomalhaut		
Instructor assigned objects		

Unit 5.8

The Apparent Motion of the Moon

OBJECTIVE

To observe and plot the apparent motion of the Moon through the constellations, and to use these data to determine the rate at which the Moon moves through the constellations.

INTRODUCTION

Because the moon is the Earth's nearest neighbor, its apparent motion is the most rapid of any celestial object except the meteors. For an observer on the surface of the Earth it is easy to observe that from full moon to full moon requires approximately 29.5 days. This period is called the synodic period or month. When this period is observed it is important to remember that the configuration of the Sun, Earth and Moon determines the starting and stopping point for the period. Thus the motion of the Earth around the Sun has an effect on the time between two full moons. If the configuration of the Earth, Moon and a distant star is used, the amount of time for the Earth and Moon to return to the same configuration is found to be approximately 27.3 days. This is the sidereal period of the Moon.

Each night, because of the rotation of the Earth, all celestial objects appear to move from East to West. The Moon, as a satellite of the Earth also appears to move relative to the stars during its period. This motion is easily noted by observing the changing phase of the Moon. As the orientation of the Sun, Moon, and Earth system changes, an observer on the surface of the Earth will see the Moon appear to move relative to the background of fixed stars each day. This motion will be in addition to the daily motion of the celestial sphere. The fact that an observer on the surface of the Earth can see different parts of the illuminated surface of the Moon is due to the relative orientation of the Sun, Earth, and Moon at different times, e.g. the quarter phases indicate that the Sun, Earth, and Moon form a right triangle while the full or new phase indicates that the Sun, Earth and Moon are in a line.

APPARATUS

A starfinder or a cross-staff.

PROCEDURE

1) Set up a starfinder to read R.A. and Dec. as described in Unit 3.3. Use a reference star close to the Moon. If you are to use the cross-staff refer to Unit 3.2.
2) Measure the R.A. and Dec. of the Moon. Record the measurement on the data table below, include the date, time and sketch of the phase of the Moon. It is a good practice to plot the measurement on the star maps before you take down the starfinder. Determine the constellations that are near your plotted point, and see if the Moon really appears to be near these constellations.
3) The data table has space for 12 measurements. The weather and/or other problems my prevent you from making all 12 measurements. The minimum number of measurements for which you will receive full credit is ______ . A measurement each night would be excellent.
4) When you have done all of your measurements plot your data on the appropriate star maps. Use no more than a small dot or two crossed lines at the measured positions. Near the measured positions sketch the phase of the Moon, and enter the date of observation.

5) Draw a **SMOOTH** curve through your measurements to indicate the apparent path of the Moon.
6) After you have completed the plot, take the two points that are closest together in time and measure the **POINT TO POINT** distance between them. This distance may be converted to angular measure by laying the distance along the scale of your star maps. To determine the rate of motion, divide the angle between the two points by the time elasped between the two measurements. Remember that if the elapsed time was say 24 hr 15 min you would divide by 24.25 hr to determine the rate.

ALTERNATIVE PROCEDURE

1) The coordinates of the moon's positions in RA and Declination for one full synodic month will be
   a) measured by you in the planetarium,
   b) given to you by your instructor.
2) Record the dates, times, and coordinates in the data table on p. 5–61.
3) Plot the ecliptic on pages 5–65 to 5–71 as a smooth, dashed curve. The coordinates of the ecliptic are tabulated on p. 9–28.
4) Now plot the moon's coordinates on the star maps listed above. Make a small dot or cross, enter the date, and connect the dots by a smooth curve, in color if you wish.
5) You have plotted the moon's positions for a selected number of dates only. Now make tickmarks on the smooth moon curve between these dates where the moon will be every day and enter the (interpolated) dates.
6) Observe the moon by naked-eye sight. Begin two days after new moon and end a few days before the next new moon.
   a) convince yourself that the moon is indeed close to the smooth curve and near the plotted or interpolated date.
   b) sketch the phase of the moon as you see it—relative to the constellations—and draw it carefully on the smooth moon curve for the date of observation.
7) Glue or tape pages 5–65 to 5–71 together in the correct sequence.
8) For the Rate Determination consult point 6) of the regular procedure above, and use date 1 and date 2 as specified by your instructor.

Unit 5.8 Name ______________________________

Date due ______________ Instructor ______________________ Section ____________ Seat ______

DATA

Date	Time	R.A.	Dec	Phase

Date	Time	R.A.	Dec	Phase

Rate Determination

1) Points Used Date 1 ______________________ Date 2 ______________

2) Time interval between Date 1 and Date 2 ______________________

3) Angle in degrees between Date 1 and Date 2 ______________________

4) Hourly rate of motion of the Moon ______________________ degrees/hour.

Unit 5.8 Name ______________________________

Date due ______________ Instructor ______________________ Section ____________ Seat _____

Stellar Magnitudes (apparent); all stars brighter than m = 3 are drawn in

- ● brighter than m = 1.0
- ● fainter than m = 1.1 and brighter than m = 2.5
- • fainter than m = 2.6 and brighter than m = 4.0
- • fainter than m = 4.0

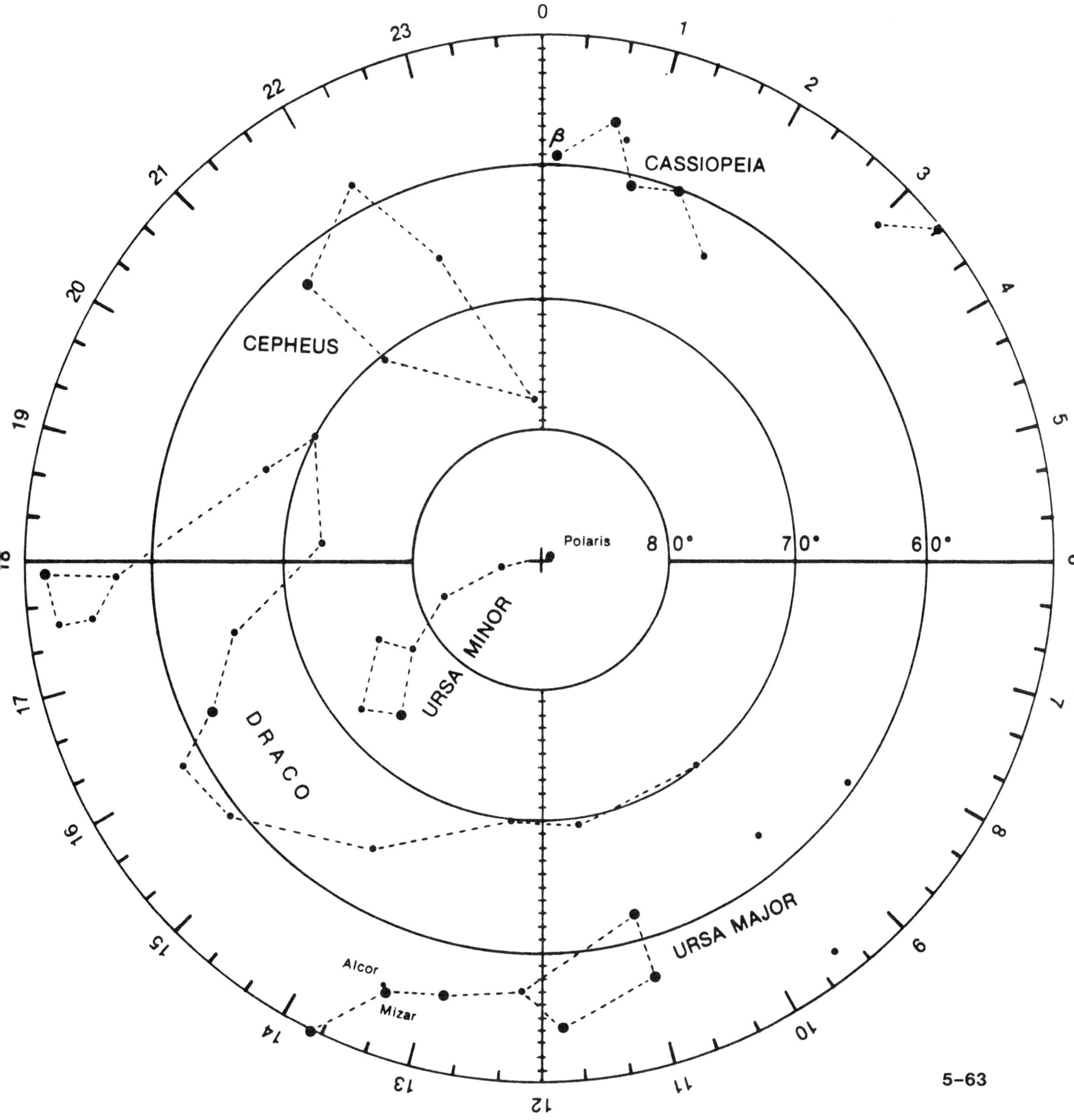

Unit 5.8 Name ______________________________

Date due ____________ Instructor ______________________ Section ____________ Seat ______

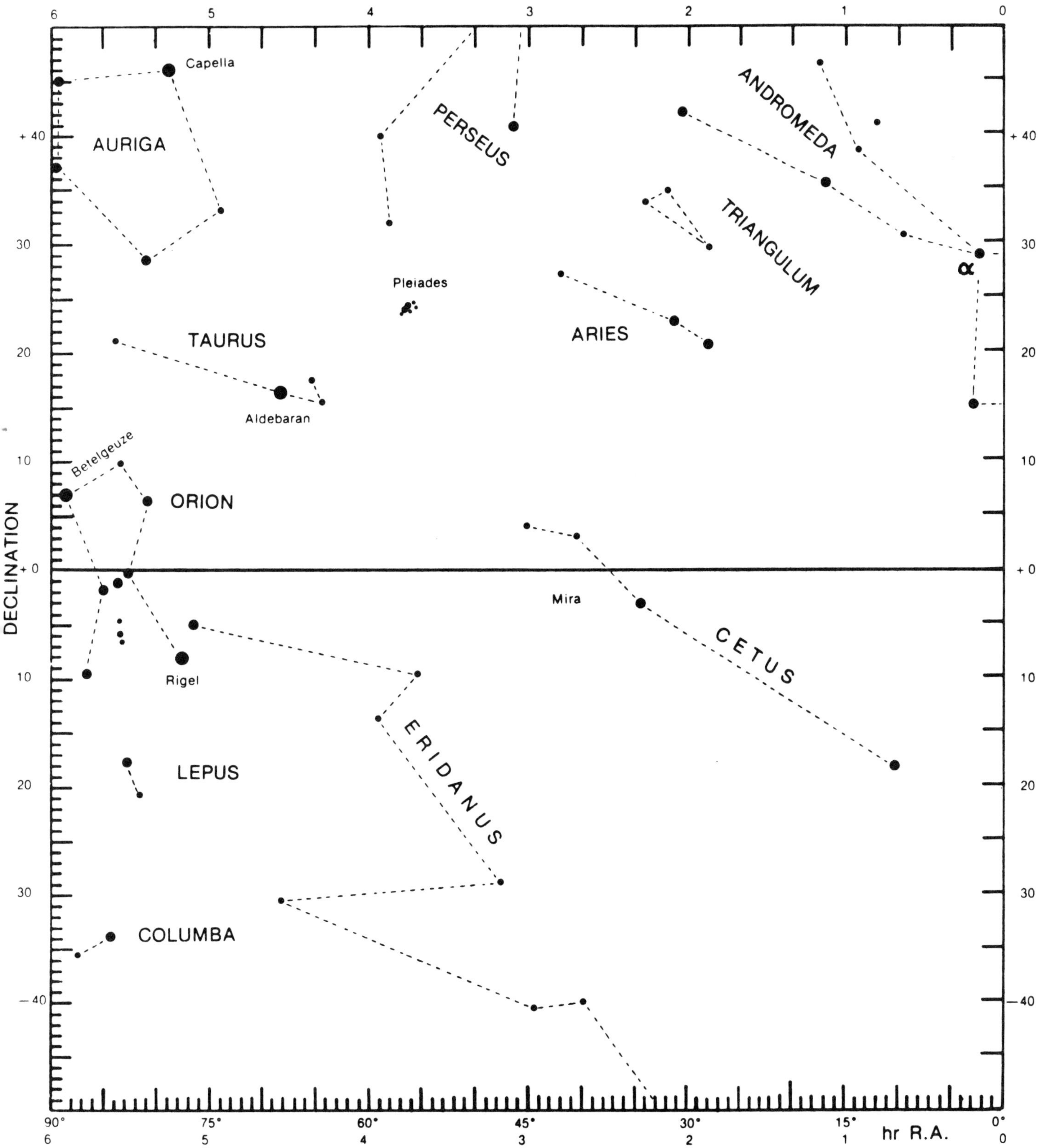

Unit 5.8 Name ____________________

Date due __________ Instructor ____________________ Section __________ Seat ______

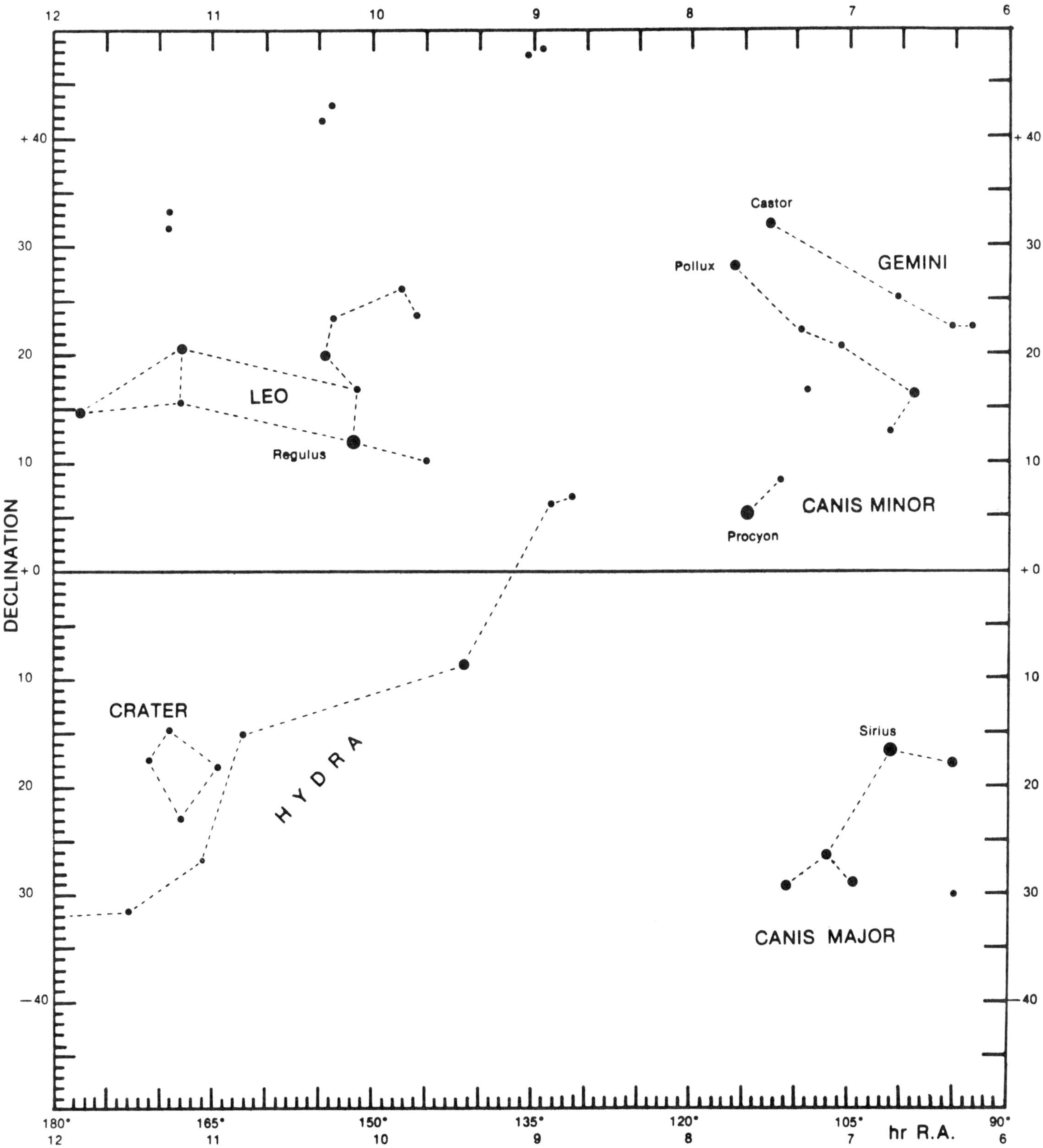

Unit 5.8 Name ____________________

Date due __________ Instructor ____________________ Section __________ Seat ______

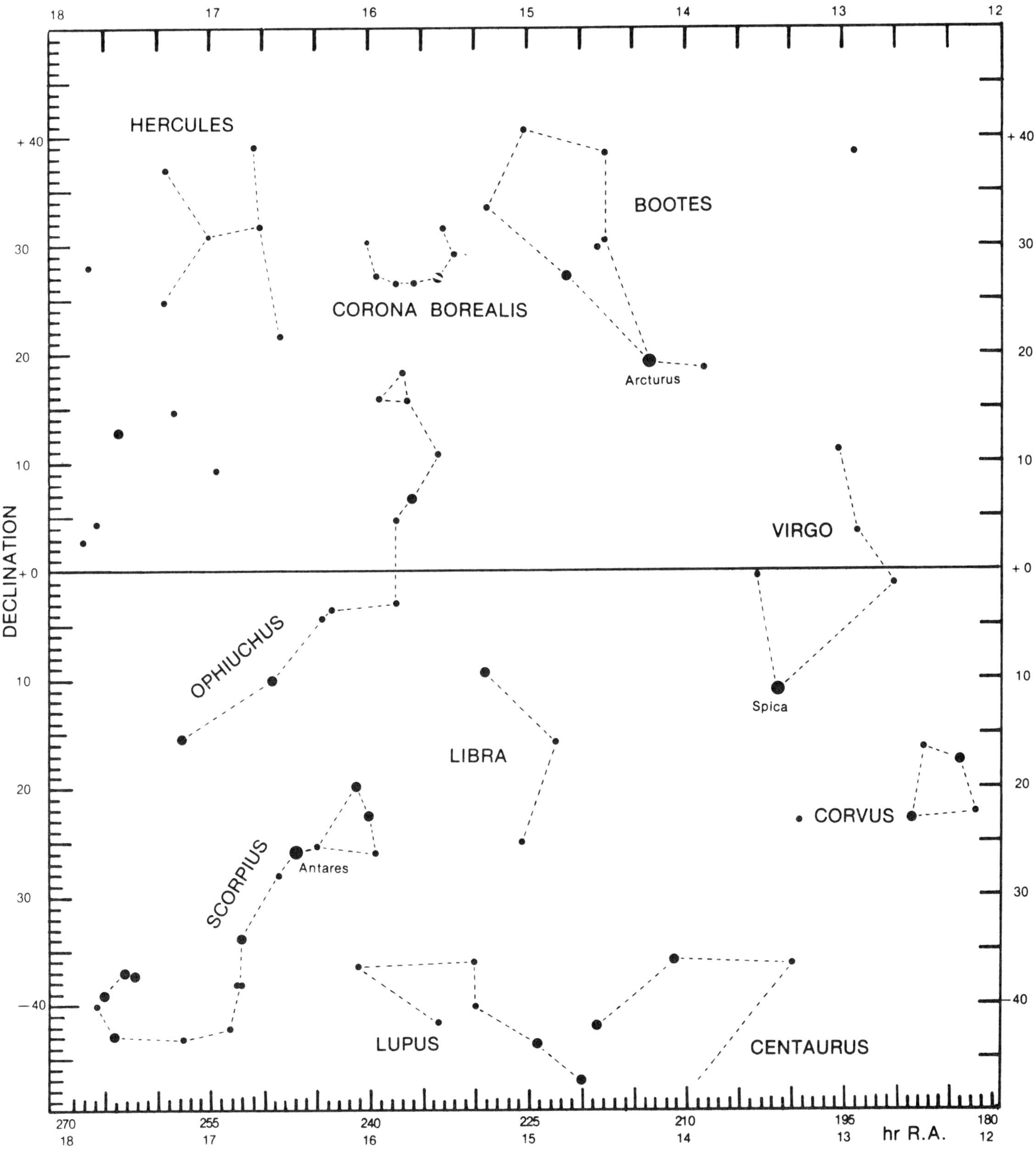

Unit 5.8 Name ____________________

Date due ____________ Instructor ____________________ Section ____________ Seat ______

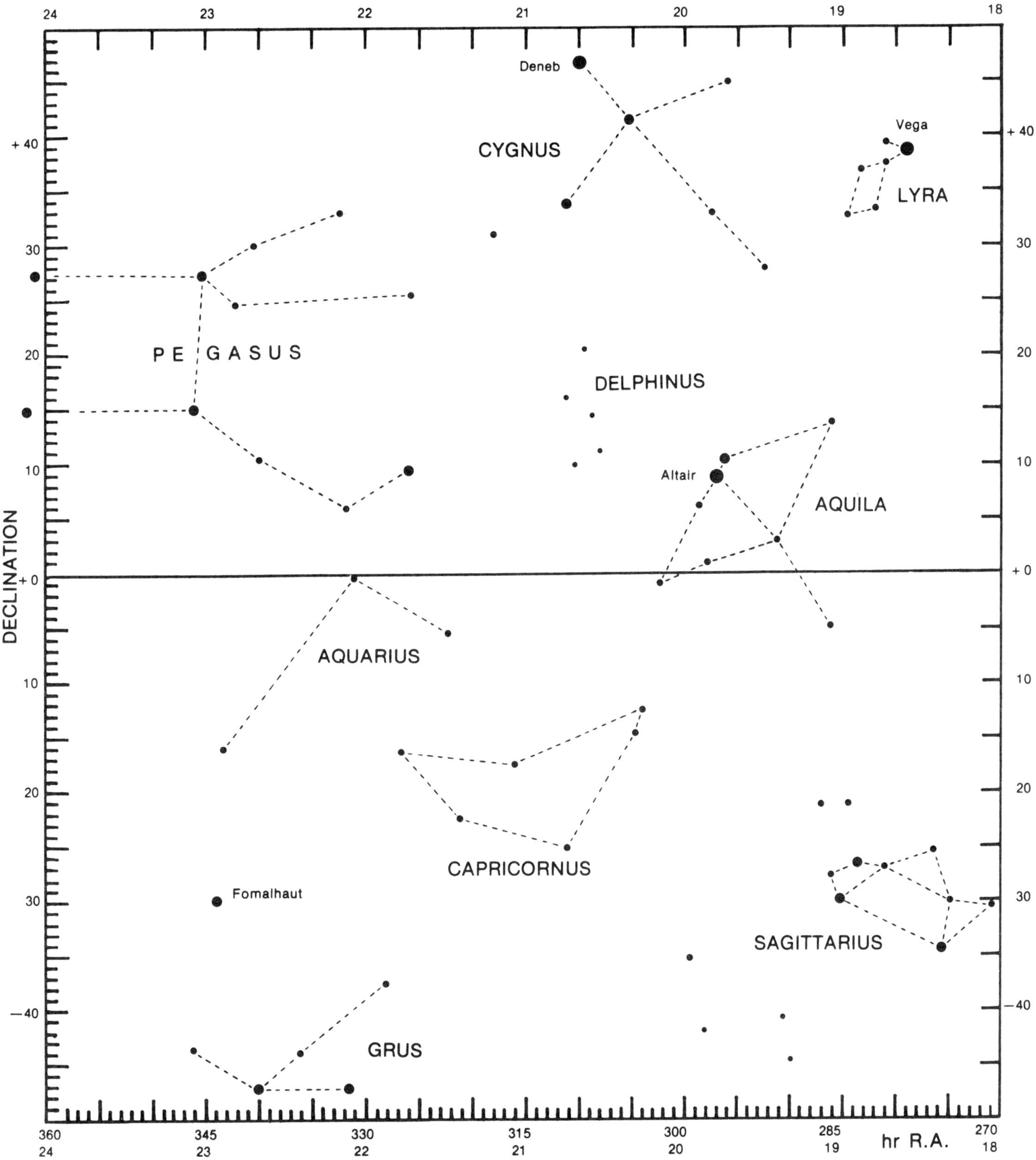

Unit 5.9

The Apparent Motion of Planets

OBJECTIVE

To observe and plot A) the motion of planets relative to the fixed stars and B) the apparent motion of Venus relative to the horizon.

INTRODUCTION

The motion of the Earth and the motion of the planets around the Sun causes the planets to appear to move in an interesting manner relative to the background of fixed stars. If the solar system is observed from a point in the direction of Polaris, all planets would appear to revolve in a counterclockwise direction. For the observer on the Earth, however, this motion appears generally as a West to East motion relative to the distant stars. As a consequence of Kepler's third law the periods of the inner planets will be shorter than Earth's (1 year) while periods of the outer planets will be longer. If you consider the motions of all planets the model indicates that planets closer to the Sun than Earth will catch and pass the Earth during some period of time. The Earth in turn will catch and pass the planets that are a greater distance from the Sun than the Earth. The result of these motions is that the planets appear to change their positions in a very interesting manner.

The inner or inferior planets will appear to the Earth observer as objects that are always near the Sun. Because of the relative positions of the Earth, Sun and inferior planets, these objects are seen only near Sun set or Sun rise. Venus is a very commonly observed object and is easy to see with the unaided eye. Venus is the object often referred to as the morning or evening star. Because of its brightness it is seen first in the evening and can be observed after Sun rise in the morning. Mercury is such a small object, and so near the Sun that observation must be made just after Sunset or Sunrise. However, if the apparent position of either of these planets is measured and plotted on your star maps, the apparent motion would be West to East except during that time when the planet is overtaking the Earth. As the inferior planets overtake the Earth they appear to reverse the West to East apparent motion and move East to West. That appears to the Earth observer as an oscillation back and forth across the Sun from the evening sky to the morning sky. When the apparent motion of any planet is East to West the motion is called retrograde.

For the outer or superior planets the motion will appear quite different. In this case the Earth will be overtaking these planets and the observed motion will appear to slow in the West to East direction, stop, move East to West forming a loop, and begin the West to East motion again. During the East to West movement, the planet is said to be in retrograde motion.

PROCEDURE

A) To observe the apparent motion of any planet, set up the starfinder to measure R.A. and Dec (See Unit 3.3). Use the following data table on page 77–9 to record the positions of the planet. The data table has spaces for twelve measurements. The motion of the more distant planets will be much slower than the motion of say Mars or Venus. If the motion appears to be very slow your measurements may be taken every other week for several months. After you have taken your data, plot these data on the appropriate star maps.

B) Venus is one of the easiest and most interesting planets to observe. For this project you should make the observation as outlined in Procedure A and also measure the altitude and azimuth of the object. Set up your Starfinder as described in Unit 3.3. However, the altitude-azimuth must be measured at the **SAME TIME** of day each time. Use the following table on page 5–81 to record your data and plot the altitude and azimuth measurements on the graph paper supplied with this unit. Put the date of observation near the points on the altitude-azimuth graph. You will note that the scales are already drawn on the graph paper. If you begin your measurements when Venus is approaching the horizon you may wish to change the altitude scale to allow you to display more points for a shorter period of time.

Unit 5.9 Name ____________________

Date due ____________ Instructor ____________________ Section ____________ Seat ______

Altitude-Azimuth of Venus

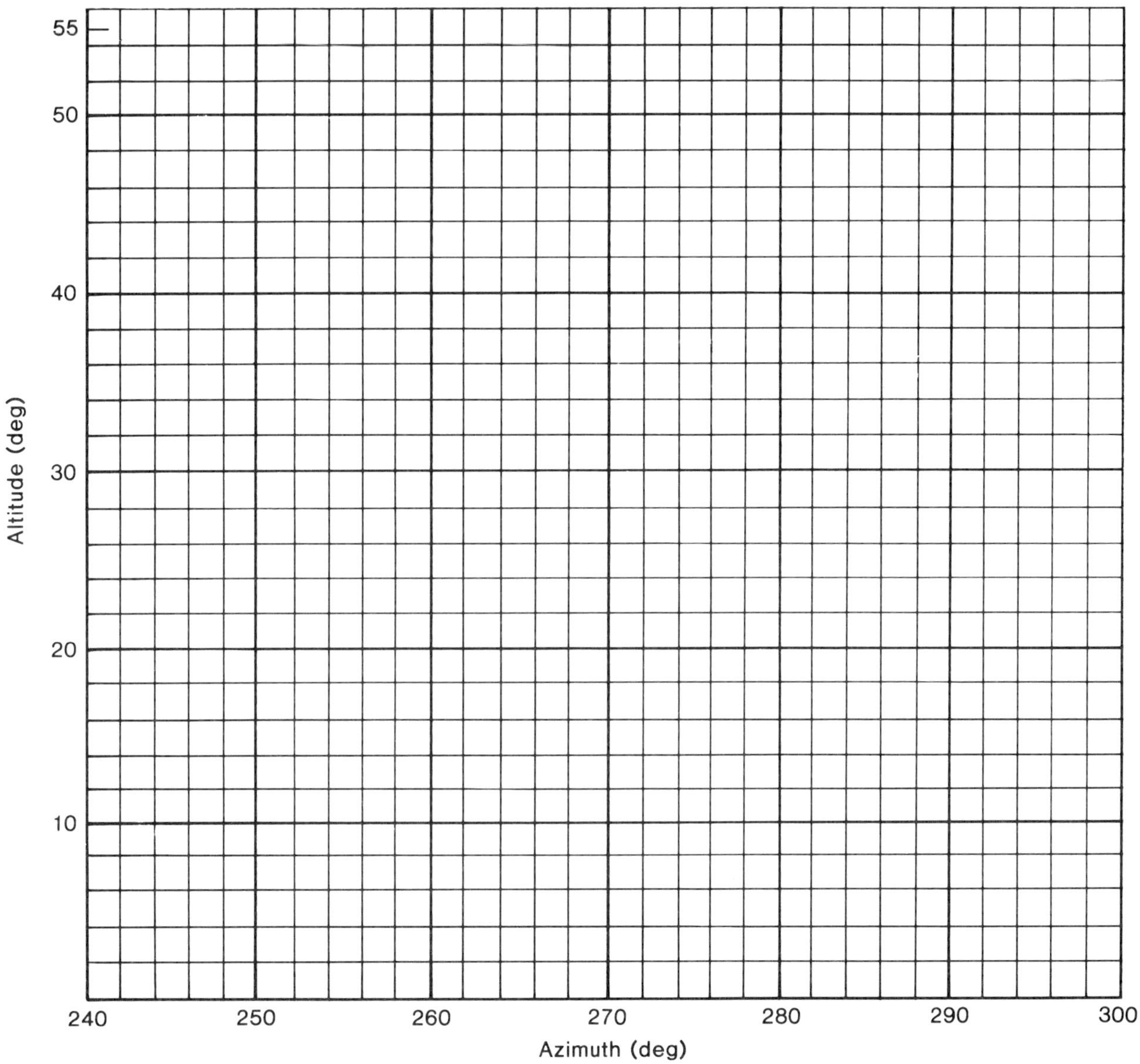

Unit 5.9 Name ______________________________

Date due ______________ Instructor ______________________ Section ____________ Seat _____

Planet ______________________________

Date	Time	R.A.	Dec

Unit 5.9 Name ______________________________

Date due ______________ Instructor ______________________ Section __________ Seat ____

Planet ______________________________

Date	Time	R.A.	Dec

Unit 5.9 Name ______________________________

Date due ____________ Instructor ______________________ Section ________ Seat ____

Venus Time of observation ______________

Date	Azimuth	Altitude

6

THE SOLAR SYSTEM

Unit 6.1

A Scale Model of the Earth

OBJECTIVE To draw models of the Earth to scale, showing its main features.

PROCEDURE Collect pertinent information from your textbook, from an atlas, or from other sources.

a) Construct model A on p. 6–3, showing Earth's large features,
b) construct model B on p. 6–5, showing the crust and the lower atmosphere of Earth.

Model A **MUST** include all of the following information:

The inner and the outer core, the mantle, and the crust of the Earth;
The layers of the atmosphere, starting with the Mesosphere and ending with the Van Allen belts;
The main chemical composition of the Earth's inner layers;
The main constituents of Earth's outer atmospheric layers.

Model B **MUST** include all of the following information:

The thickness of the Earth's crust and its chemical composition, including the 10 most common chemical elements;
The greatest ocean depth and the highest mountain;
The atmospheric layers of the Earth up to an altitude of 100 km and their main chemical composition, including the 4 most common constituents of Earth's lowermost atmospheric layer.

Both models **MAY** include any additional information which you may find interesting, for example: Temperatures, Densities, Atmospheric Pressures, Electrified Layers, Location of Auroras, Satellite Orbits, etc.

Unit 6.1 Name ______________________________

Date due __________ Instructor __________________ Section __________ Seat ______

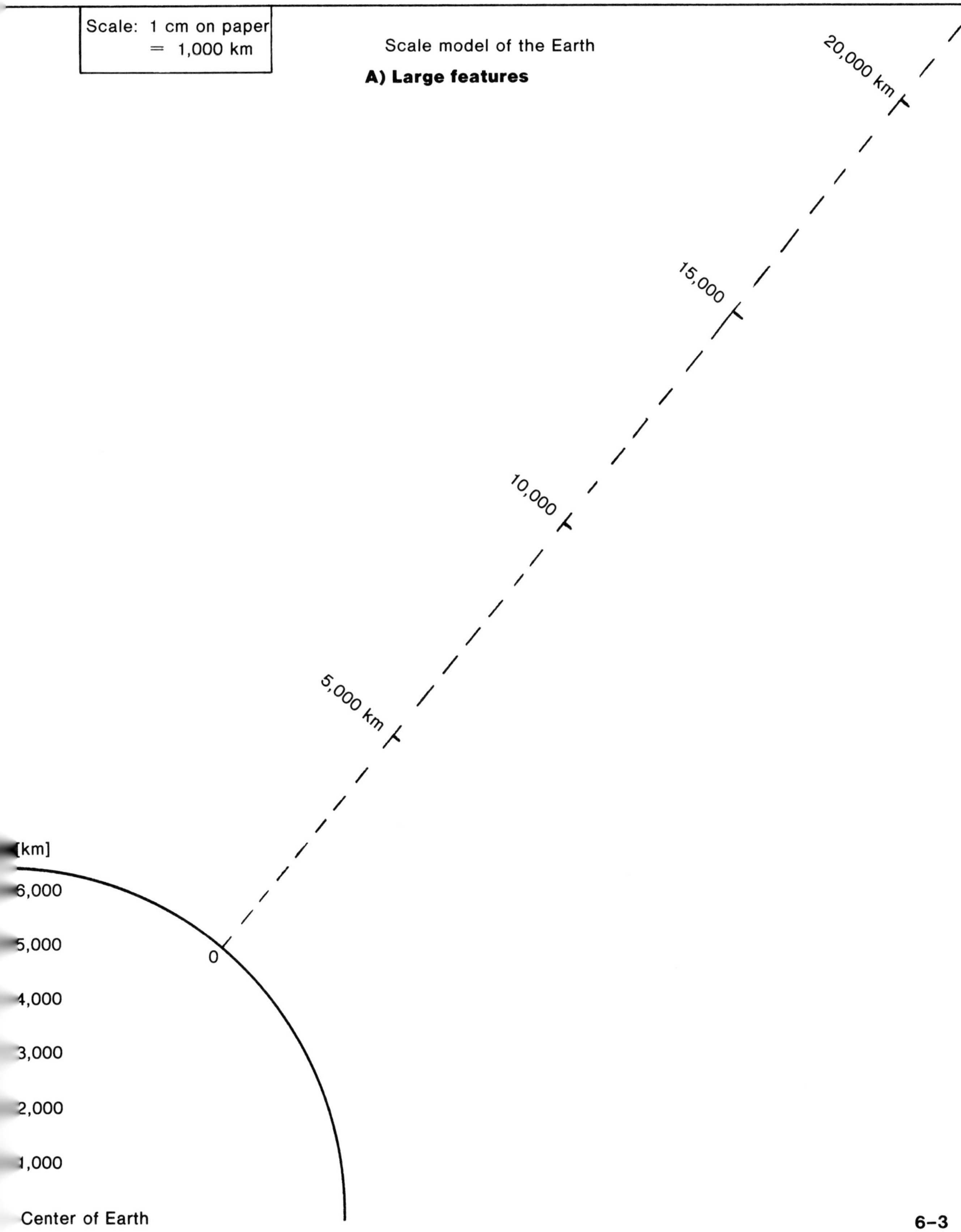

gure 1

Unit 6.1 Name ______________________________

Date due ______________ Instructor ______________________ Section ____________ Seat ______

Scale Model of the Earth

B) Crust and Lower Atmosphere

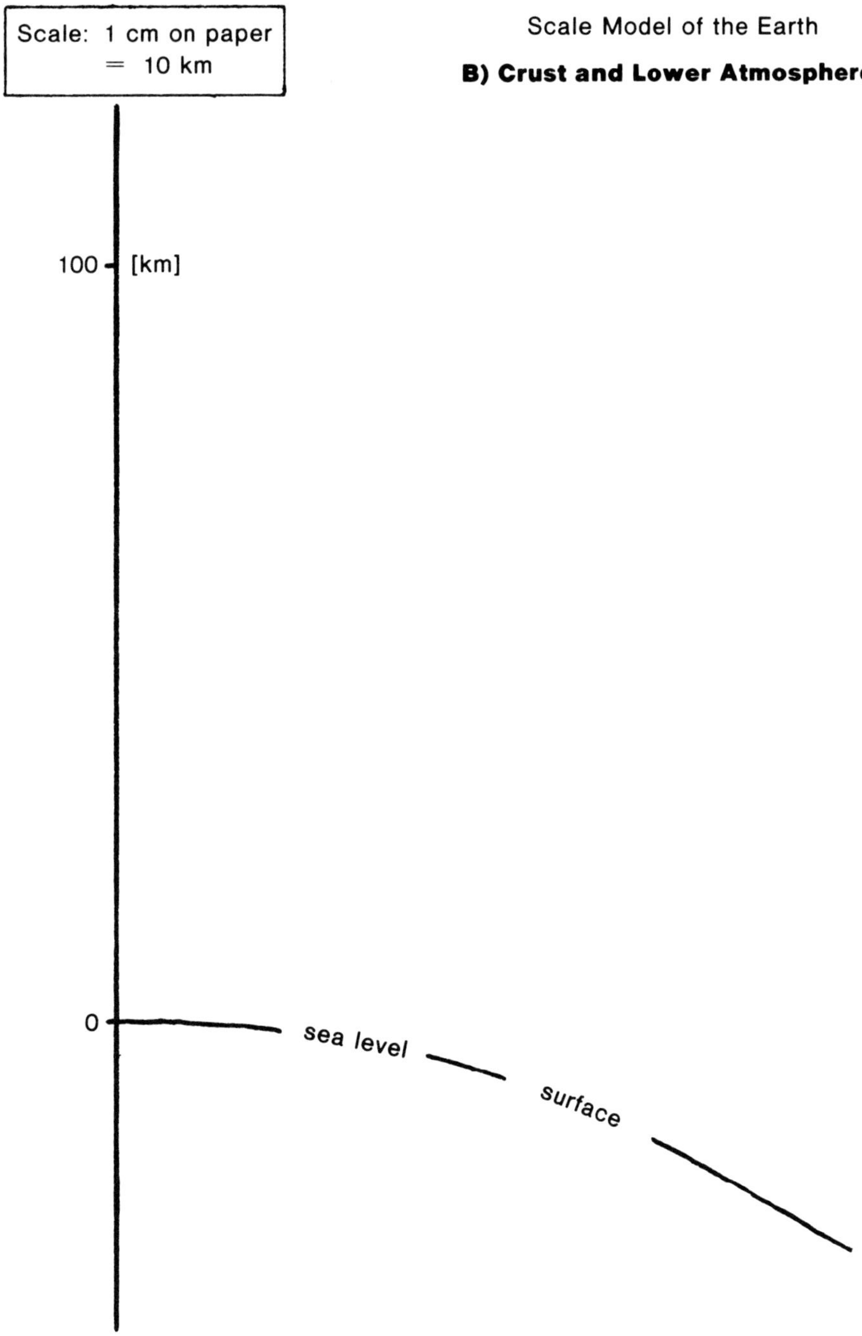

Figure 2

Unit 6.2

A Scale Model of the Solar System

OBJECTIVE

To construct a scale model of the solar system, showing the distances within the solar system, the relative sizes of its components, and some other interesting features.

INTRODUCTION

The distances within the solar system are so huge and the sizes of its components so (relatively) small that it is impractical to construct a model which uses the same scale for both distances and sizes.

For example: If you would draw the diameter of the Earth as a very small circle with 1 mm diameter (∘), Mercury would be a mere dot, the sun would be a circle with about 10 cm or 100 mm diameter, the distance sun-earth in this scale would be about 15 m, and Pluto would be about 600 m from the model sun. Since we cannot expect you to purchase that much paper and glue it together, we will construct the solar system model in three parts.

Scale A will be used for the inner planets up to 5 Astronomical Units (AU), Scale B will be used for the whole solar system, but for distances only, not sizes. Scale C will be used for the sizes of the system components only.

In addition to distances and sizes, the model **MUST** show the following features:

1) The inclination of planetary orbits to the ecliptic.
2) The N and S poles of all major planets, except Pluto.
3) The equator of all major planets, except Pluto.
4) The inclination of the equator to the orbit for all planets, except Pluto.

Sources of information: Your textbook, planetary data table in the appendix.

PROCEDURE

SCALES A AND B on p. 6–9.

These two scales are for distances of the planets from the sun only! Draw the planets as small dots at the proper distances from the sun. The solar system is seen from the side, so to speak; note the ecliptic. Draw all planets either above or below the ecliptic, using the proper angle for their orbit's inclination from the ecliptic. This can easily be done correctly for Mercury and Pluto. For all the other planets use common sense to denote their inclination angle. Designate also the location of the minor planets in the appropriate scale.

SCALE C on p. 6–11.

The rim of the sun is drawn approximately to scale on the left side of p. 6–11. The scale for model C is: 1 cm on paper = 6,400 km = radius of the earth. Use this scale factor to answer the question for the "model" radius of the sun and write on provided space in the solar rim. Draw the planets to scale, indicate the required features 2) through 4). This has been done already for the earth as an example.

Your textbook and the table in the appendix provide much more useful and interesting information. You may wish to write in this information. You can arrange the planets in model C in such a way that you have sufficient room left to do this. There is no objection to include this information inside the circle of the larger planets.

Considering the wealth of visual material, produced by contemporary space exploration, and its easy accessibility, you may also wish to expand this model of the solar system on your own.

Unit 6.2 Name ____________________

Date due ______ Instructor ______ Section ______ Seat ______

A scale model of the solar system

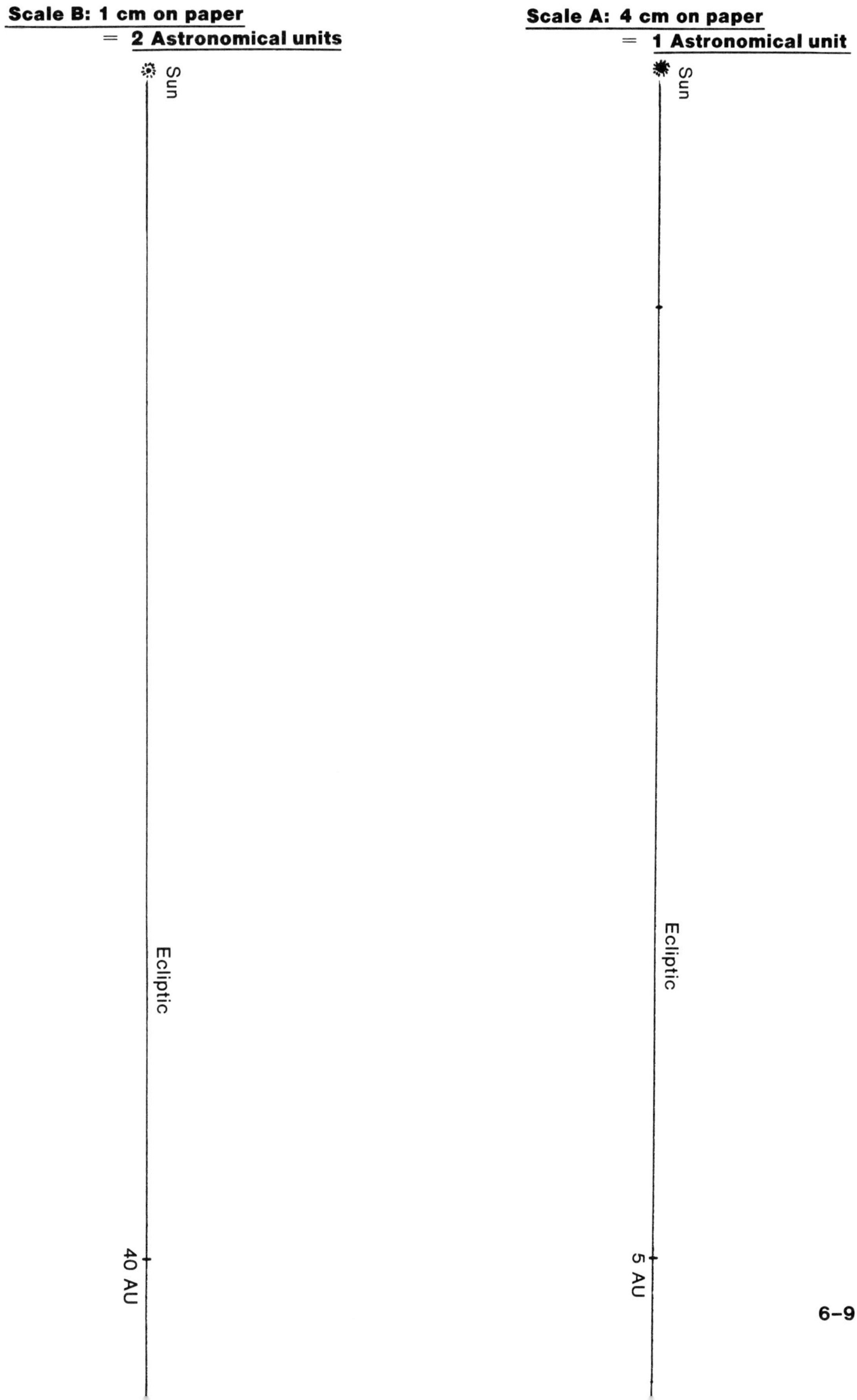

Figure 1

Unit 6.2 Name ____________________

Date due ____________ Instructor ____________________ Section ____________ Seat ______

A scale model
of the solar system

Scale C: 1 cm on paper = 6,400 km = Earth's radius

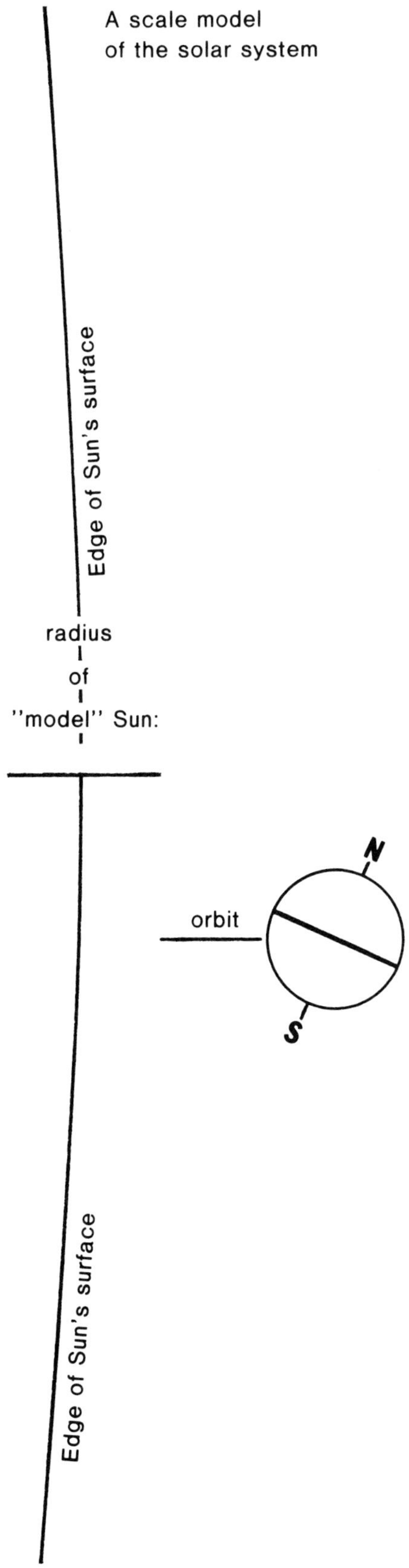

Figure 2

Unit 6.3

Planetary Motions and Ecliptic Coordinates

OBJECTIVE To study and plot the motions of planets relative to the Sun and the Earth.

INTRODUCTION At certain times of the year Venus is a "morning star," at others an "evening star". In one year many planets are visible during the night, in another year hardly any are visible. You can always consult some publications, e.g. "Sky and Telescope," to find out which planets are visible and at what times. This unit intends to help you understand these motions of planets. You may also construct your own "planetary clock".

We will consider only the planets visible to the naked eye; Mercury, Venus, Mars, Jupiter, and Saturn. As you already know, their orbits lie very nearly in the same plane as the Earth's orbit around the Sun, the **ECLIPTIC.** For the first part of this unit we will assume that these planets lie in the same plane.

Their orbits are ellipses, but not very elongated ones. In our figures we will draw the orbits as circles, introducing only minor inaccuracies.

We choose as our reference plane the ecliptic with the Sun at the center. On the beginning of spring, March 21, the Sun is in the vernal equinox (constellation Pisces). This is the origin of our

ECLIPTIC OR HELIO-CENTRIC COORDINATE SYSTEM

The helio-centric angle is counted from 0° at the vernal equinox counterclockwise. The motion of all planets is also counterclockwise, so is the direction of rotation of all planets, except Venus. When we say that the Sun is in the vernal equinox, we mean: as seen from the Earth. The vernal equinox (0° helio-centric), the Sun, and the Earth are in one line. Extending this line across the Earth to the degree scale, we read the helio-centric coordinate (longitude) of the Earth to be 180°, see Fig. 2. A quarter year later, at the summer solstice on June 22, what are Earth's helio-centric coordinates? Draw a line in Fig. 2 from the summer solstice (90° helio-centric) through the Sun and the Earth to the degree scale. The Earth's coordinates are: 270° helio-centric. Let us consider the four major positions of the Sun and Earth. Consult Fig. 2 and compare with the table below:

As Seen from the Earth, the Sun is in:	Apparent Coordinates of the Sun as Seen from Earth	Helio-Centric Coordinates of the Earth
Pisces, vernal equinox	0°	180°
Gemini, summer solstice	90°	270°
Virgo, autumnal equinox	180°	360° or 0°
Sagittarius, winter solstice	270°	90°
Pisces, vernal equinox	360° or 0°	180°

In 1 year the Earth revolves once around the Sun, or the Sun seems to move once around the Earth through the Zodiac constellations. Their names are shown outside the circles in both figures 2 and 3. They appear about equally spaced, each occupying an angular width of 30°. The actual borders of the astronomical Zodiac constellations, however, vary widely in width. Consult an astronomical atlas for exact borders of the constellations.

Assume that on June 22 the planet Mars is in opposition with the Earth. Sun, Earth, and Mars are in a straight line. What are the ecliptic coordinates of these three bodies? The Sun (as seen from the Earth) is in the summer solstice, at 90° helio-centric. The Earth at 270° helio-centric, and Mars also at 270°. If we know the helio-centric coordinates of the planets for any given date, then we can plot their successive positions in our ecliptic coordinate system, because we know their sidereal periods. See Appendix, table 2, p. 9–21 for their values. In Fig. 2 the Earth's position is indicated in its orbit for every first and fifteenth of each month; in Fig. 3 for every first of each month. In Fig. 2, the orbits of Mercury, Venus, Earth, and Mars are drawn to scale, the orbits of Jupiter and Saturn are not drawn to scale. In Fig. 3, the innermost orbit is the Earth's, Mars's orbit is omitted, and the orbits of Jupiter and Saturn are drawn to scale. Now let us see what we can do:

EXAMPLE 1

WITH A PENCIL, indicate the Earth's position at 270° helio-centric. Draw a line from the Earth's position tangent to Mercury's orbit, (towards the vernal equinox).

QUESTIONS AND ANSWERS

a) The Sun appears in which constellation at what coordinate? In Gemini at 90°.
b) What is Mercury's **ELONGATION,** the maximum angle from the Sun? Measure this angle with a protractor: about 23°. Note: Here we **DO** encounter some inaccuracy: Mercury's orbit is much more noticeably elliptical than the orbits of the other planets; Mercury's orbit is also more inclined to the ecliptic (7°) than the others. Both facts distort the measurement you made. The actual elongation (maximum) is about 28°.
c) Is Mercury a "morning star" or an "evening star"? Consider the counter-clockwise rotation of the Earth around its axis. Mercury is visible before the Sun has risen, it is therefore a morning object.
d) The Sun will rise how many hours and minutes after Mercury has risen? 23° = 15° + 8°. 15° is equivalent to 1 hr; 8° is equivalent to 32 min. Answer: 1 hr 32 min (see the appendix, p. 9–17 for conversion of angular measure into time measure).
e) What is the helio-centric coordinate of Mercury for this case of maximum elongation? Draw a line from the sun through the point on Mercury's orbit where the tangent touched it, and continue this line to the degree scale. You read: about 335°.

This is a reverse case. Normally you would start with given coordinates and then figure out where the planet appears to be.

EXAMPLE 2

Use pencil lines again! Leave the Earth at 270° helio-centric. Put Mars in its orbit with helio-centric coordinate 222°. Draw a line from the Earth to Mars. Note that this line is tangent on the Earth's orbit, the so-called "quadrature".

QUESTION AND ANSWERS

a) If you were on the planet Mars, what would be the elongation of the Earth? The angle Sun-Mars-Earth as measured by a protractor is about 40°. As seen from Mars, the Earth would be a "morning star", rising 2 hours 40 minutes before the Sun would rise above the horizon of Mars.

b) Is Mars visible in the morning before sun-rise or during the night? Draw a larger circle for the Earth, indicate the Earth's direction of rotation, then look: When the Sun is on Earth's horizon and setting, Mars would be high in the sky, in fact, it would go through the meridian. It will be visible for about 6 hours before it sets in the West, because it is 90° away from the Sun (90° East of the Sun).

c) In which constellation would Mars be? Mars is 90° East of the Sun. The Sun itself is in the summer solstice, in the constellation Gemini. This question **CANNOT** be answered by drawing a line from the Earth to Mars and extending this line to the circle of Zodiac constellations. If you would do that, the answer would be Libra—which is wrong. Why? Fig. 2 and Fig. 3 are drawn to scale for some planets as stated. They are grossly out of scale to accommodate any stars. Any of the stars forming any of the Zodiac (or other) constellations are so distant that the Zodiac constellations would have to be drawn around a circle of some tens of miles radius.

The other possibility—maybe easier to imagine—is the following: Suppose the stars forming the Zodiac constellations are where they are drawn in the 2 figures. Then the whole solar system would be compressed almost into a point—as seen from the stars. The Earth would touch the Sun, Mars would be slightly more outward, but still 90° away from the Sun as seen from the Earth. Mars would therefore appear in the constellation Virgo, see the small inset in Fig. 1. So, how do you find the answer to the question: In which Zodiac constellation will a certain planet be?

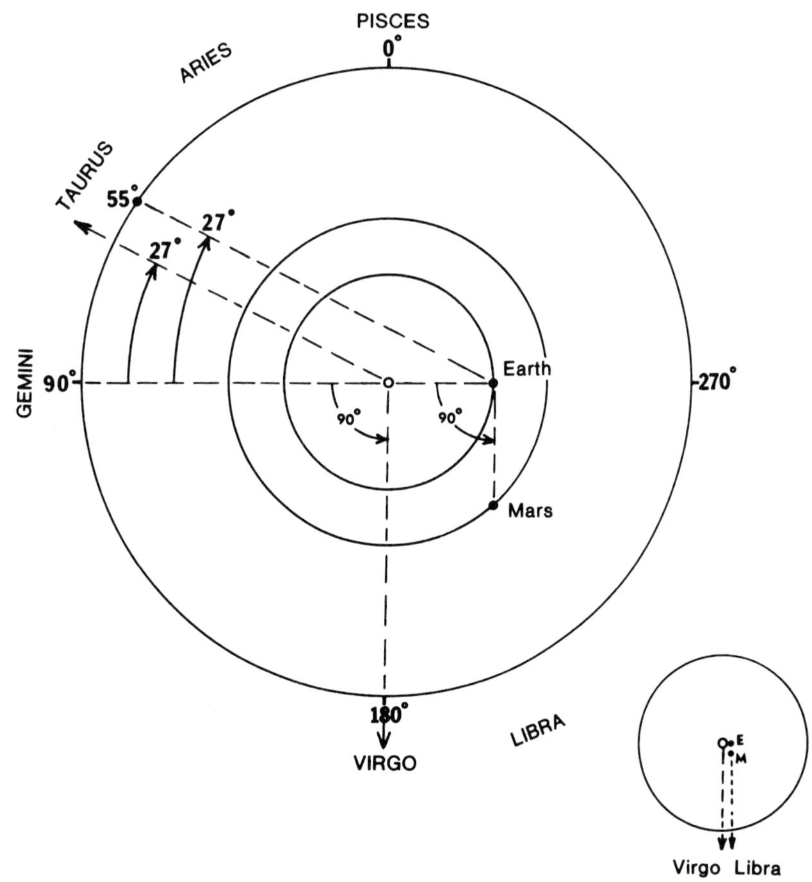

Figure 1

Fig. 1 gives the general answer: Continue with Mars. You have drawn the line from Earth (270° helio-centric) to Mars (222° helio-centric). The angle between the Sun and Mars is 90° **AT THE EARTH.** Now draw a line from the Sun with angle of 90° **AT THE SUN,** or a line parallel to the line Earth-Mars going through the Sun. Where this line meets the Zodiac circle is the position of Mars in a given Zodiac constellation, in this case Virgo. To make you more familiar with this kind of construction: Consider Asteroid X. Its orbit is drawn to scale and X has the heliocentric longitude of 55°. It is 27° from the Sun as seen from the Earth and measured by a protractor. In which constellation will it appear to be? A constellation 27° from the Sun on the Zodiac belt, in this case Taurus.

PROCEDURE

1) Use one of the Fig. 2 and 3 for your own purposes. You may, for example, mount the figures on cardboard or soft wood. Use pins for the planets, divide the planetary orbits into the same time intervals as earth's orbit has been divided, and move the planets accordingly. You need, of course, the initial coordinates for the planets. You can get that information from: "The American Ephemeris and Nautical Almanac". The helio-centric coordinates (longitude only) are given on p. 6–18, line 1) for January 1, 1980 for the naked-eye planets. Use them as instructed.
2) As a project for more advanced students: Refine the drawings in Fig. 2 in two ways.
   a) Draw Mercury's orbit as an ellipse. You can use either Mercury's perihelion and aphelion distances or you can use its mean distance in conjunction with the eccentricity of the orbit. You also need to include in this refinement the orientation of the major axis of Mercury's orbit. This orientation changes, by the way, and is the famous "Perihelion Motion" of Mercury. You should now be able to measure Mercury's maximum elongation much more accurately. (A still better refinement would be the substitution of Earth's actual elliptical orbit for the drawn circular one.)
   b) As stated at the beginning of this unit, we neglected the inclination of the planetary orbits to the ecliptic, and assumed that all planets revolve around the Sun in the plane of the ecliptic, i.e. the plane of paper in Figs. 2 and 3. At certain times the planetary orbits are above the ecliptic, at others below. In the drawings you could denote planetary orbits above the ecliptic by solid lines, and the parts of their orbits below the ecliptic by dashed lines, for example. How could you find that information? From tables which give the RA and Declination of planets, e.g. "The American Ephemeris And Nautical Almanac". If the Declination values for a planet are larger than the values for the ecliptic, then the orbit is above the ecliptic as long as the values stay larger. There are two points for which the Declination values for the planet are the same as those for the ecliptic (path of Sun). At these two points the planetary orbit goes through the plane of the ecliptic. If the Declination values for the planet are smaller than the Declination values for the ecliptic, the orbit is below the ecliptic.

PROJECT 1

Use the coordinates (longitudes) for the planets given on line 1) on p. 6–18 or the coordinates to be specified by your instructor. Plot the positions of

a) Mercury   b) Venus   c) Earth   d) Mars   e) Jupiter   f) Saturn

g) ____________   h) ____________   for the date: ________ .

Answer the following questions on an extra sheet of paper:

1) Which of the plotted planets are visible after sunset during the night?
2) Which of the plotted planets are visible before sunrise as ''morning planets''. Take ''before sunrise'' to mean from 4 hours before to sunrise.
3) Which of the planets are ''evening planets'', visible after sunset to about midnight?
4) Which planets are not at all visible during the night?
5) Match all assigned planets with the constellations they appear to be in for the specified day.

6) ______________________________

7) ______________________________

8) ______________________________

9) ______________________________

10) ______________________________

11) ______________________________

12) ______________________________

13) ______________________________

PROJECT 2

You can use either Fig. 2 or Fig. 3 to demonstrate the retrograde motion of planets.

Consider the Earth and Mars in opposition. Draw a line from the Earth through Mars towards the stellar background. Mars will appear in a certain constellation. Now move both planets for the same time intervals, for example one week, along their orbits. Draw a line from Earth through Mars for everyone of these positions. You will note that during certain times Mars moves in one direction relative to the stellar background (as seen from Earth along the line of sight); at other times Mars reverses its direction of motion. If you do that until Mars is again in opposition, you can determine the so-called **''SYNODIC PERIOD''** of Mars. Extra sheets with Figs. 2 and 3 are provided in the appendix.

Similarly for the other planets.

Ecliptic or Helio-Centric Coordinates:
In Degrees, Longitude Only, Rounded Off to Nearest Full Degree

	apparent Sun	Mercury	Venus	Earth	Mars	Jupiter	Saturn	Uranus	Date
1	280°	242°	358°	100°	131°	151°	171°	232°	1/1/1980
2									
3									
4									
5									
6									
7									
8									
9									
10									
11									
12									
13									
14									
15									
16									
17									
18									
19									
20									
21									
22									
23									
24									
25									
26									
27									
28									
29									
30									
31									
32									

Unit 6.3 Name ____________________

Date due ____________ Instructor ____________________ Section ____________ Seat ______

Ecliptic or Helio-centric coordinates
The orbits of Mercury, Venus, Earth, and Mars are drawn to scale.

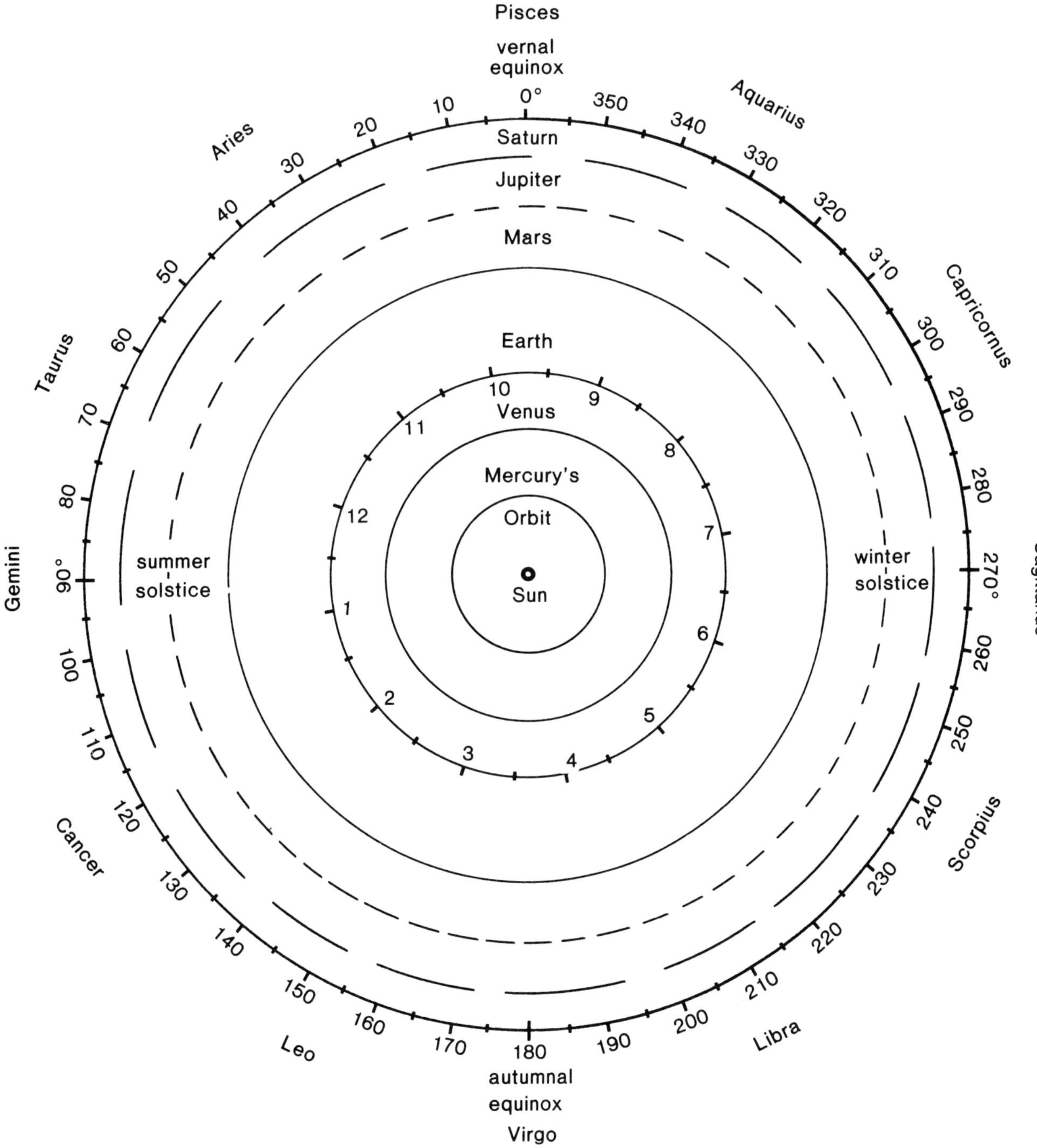

Figure 2

Unit 6.3 Name ______

Date due ______ Instructor ______ Section ______ Seat ______

Ecliptic or Helio-centric Coordinates.
The orbits of Earth, Jupiter, and Saturn are drawn to scale.

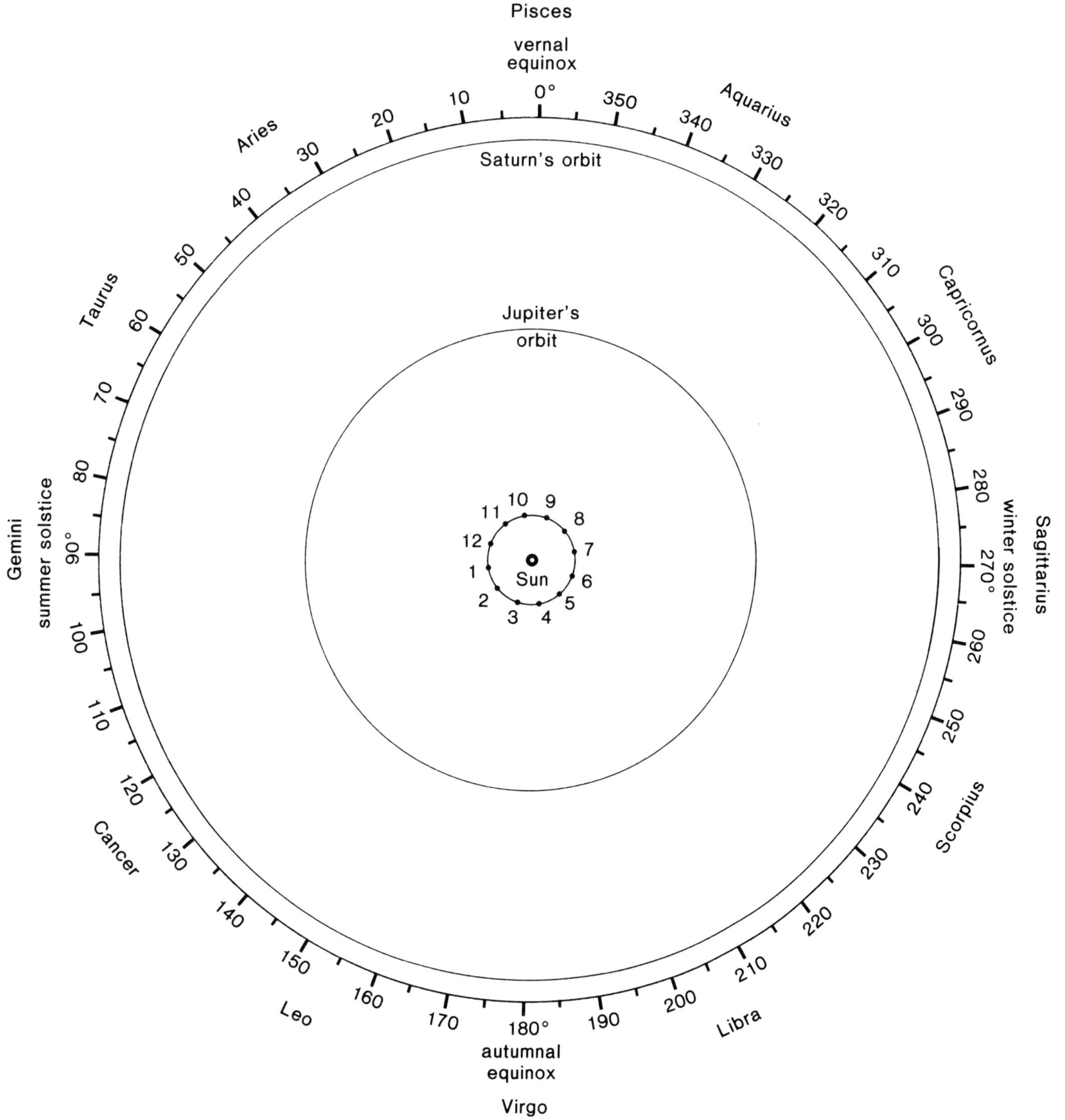

Figure 3

7

FROM STARS TO GALAXIES

Unit 7.1

Angles and Parallax

OBJECTIVE

To become familiar with angular measures, their application in a parallax measurement, and to relate them to the helio-centric parallax.

THEORY

A) ANGLES

An angular measure, very often used in astronomy, is the **"RADIAN".** An angle θ in radians [rad] is defined as:

$$\theta = \frac{\text{arc}}{\text{radius}} \text{ [rad]}$$

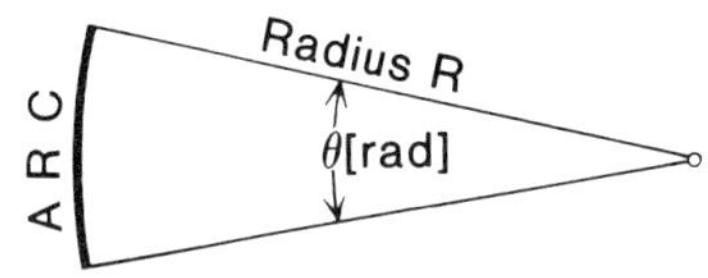

Figure 1

To establish the relation of radian to degree measure, consider the following two examples: Imagine you are standing at the center of a circle of radius R, Fig. 2:

1) Look once completely around its circumference. Your eyes sweep through an angle of 360°. What is the corresponding angle in radians?: The arc is the circumference of the circle, equal to $2\pi R$. The angle in radians is now:

$$\theta = \frac{\text{arc}}{\text{radius}} = \frac{2\pi R}{R} = 2\pi \text{ [rad]}$$

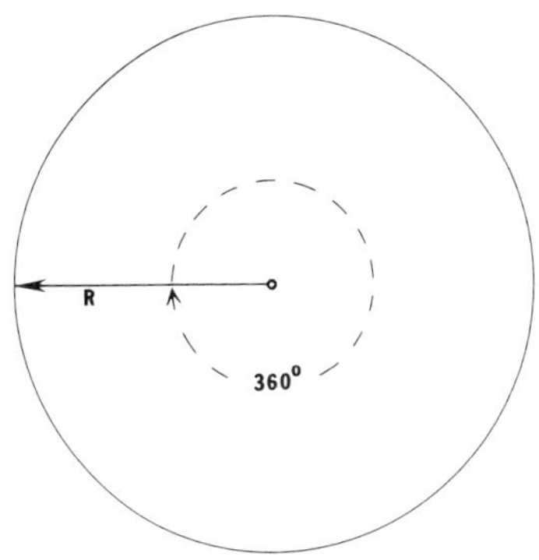

Figure 2

2) Instead of looking once completely around the circumference of the circle, look only one quarter its circumference, equal to $2\pi R/4$, or through an angle of 90°, Fig. 3. The angle in radians is then:

$$\theta = \frac{\text{arc}}{\text{radius}} = \frac{2\pi R}{4\ R} = \frac{\pi}{2} \text{ [rad]}$$

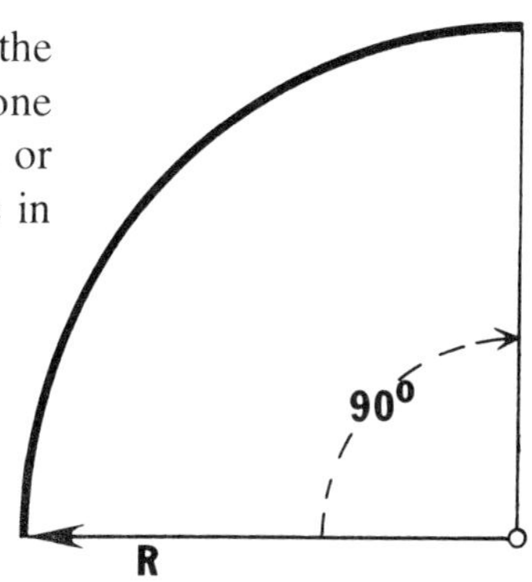

Figure 3

To convert angles from degrees (°), minutes of arc (′), and seconds of arc (″) into radians and vice versa, use the table below:

angle in ° ′ ″	angle in radian	angle in radian	angle in ° ′ ″
360°	$2\pi = 6.283$	2π	360°
180°	$\pi = 3.142$	1	57.3°
90°	$\pi/2 = 1.571$	0.1	5.73°
45°	$\pi/4 = 0.785$	0.01	0.573° = 34.4′
1°	1.745×10^{-2}	10^{-3}	3.44′
1′	2.91×10^{-4}	10^{-4}	0.344′ = 21″
1″	4.85×10^{-6}	10^{-6}	0.21″

APPARATUS

1 piece of non-stretchable string, 10 m long with a knot or other mark at the 5 m interval.
A ruler, at least 1 foot long, preferably with an inch- as well as a cm scale.
A piece of chalk or other marking device.

PROCEDURE

1) Answer question 1) on pg. 7–7.
2) Place the 10 m string straight on the ground and determine how many of **YOUR** feet are equivalent to a distance of 10 m. Answer question 2) on p. 7–7.

3) Go to the flagpoles west of the library (or to ________________________).

____

For the following measurements, use the southern flagpole only, (or use

____).

Use the 10 m string, attach it to the pole, and go 10 m away from the pole to position (1), so that **YOU,** the pole, and the southern corner of Texas Hall are aligned, (or so that you, the pole and ________________________

________________________ are aligned).

Mark your position (1).
Study Fig. 4 below! (This Fig. 4 is repeated on the question sheet, p. 7–7).
Now walk in a circular arc with radius 10 m (by using your string) towards the

south (or towards ________________), until you, the pole, and the

northern front corner of Texas Hall are aligned, (or until you, the pole, and

____ are aligned).

Mark this position (2).
Measure the length of the arc from (1) to (2) any way you like, but to the nearest 5 cm or 2 inch. If you have determined the length of **YOUR** feet, simply count the number of feet and their fraction by walking along the arc.

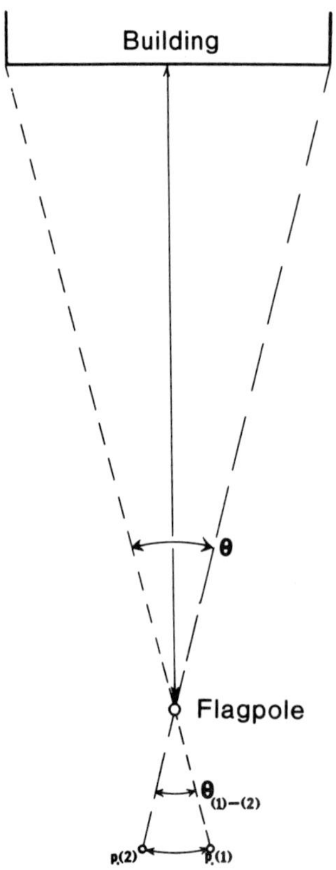

Figure 4

Place the measured value in table 3 on p. 7–7 and calculate the angle θ in radian and degree, using

$$\theta = \frac{\text{arc}}{\text{radius}} \text{ [rad]}$$

4) You have now determined the angle $\theta_{(1)\text{-}(2)}$, see Fig. 4. Since this angle is equal to the angle θ (as opposite angle) in Fig. 4 you can now calculate the

distance R from the pole to Texas Hall, (or from the pole to __________). Use the above equation again and solve it for R:

$$R = \frac{\text{arc}}{\theta} \text{[m] if the arc was measured in m.}$$

The arc you use for this calculation is the width of Texas Hall, 45 m.

(or ____)
Answer question 4 on p. 7–7.

5) Do the same with the 5 m string, i.e. distance pole to your respective positions (1) and (2) = 5 m. Place your measured values in table 5 on p. 7–7. Calculate the angles and answer question 5), again using the width of the building with the same value.

THEORY

B) THE HELIO-CENTRIC PARALLAX

Imagine you were on a nearby star and were looking back towards the solar system. The radius of the earth's orbit around the sun would then subtend a certain angle at your eye. This angle is very small. In the drawing in Fig. 5 below it is grossly exaggerated. The radius of the earth's orbit is drawn in the figure. It is 93 million mi or 1 Astronomical Unit (AU). The nearest star is about 300,000 AU from the sun! The actual angle is always less than 1 sec of arc. The helio-centric parallax is defined as:

The angle in seconds of arc at which the radius of the earth's orbit around the sun appears, as seen from a star. It is expressed in "parallax-seconds" or parsec (pc).

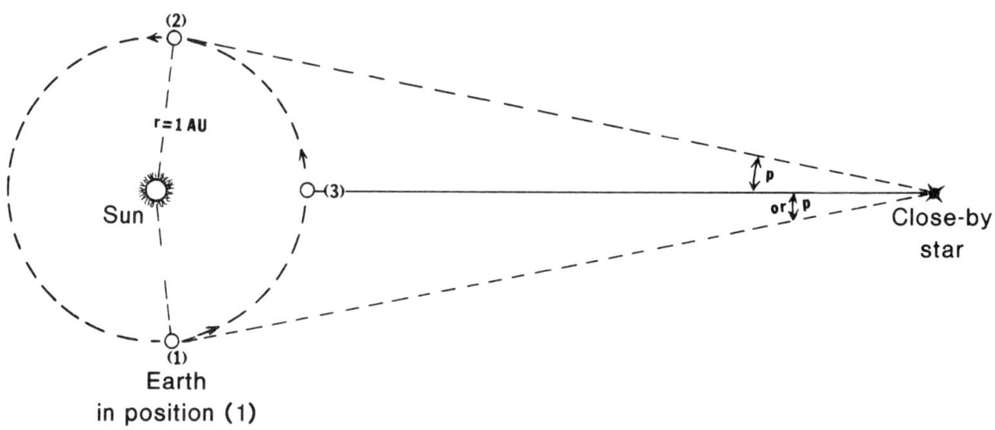

Figure 5

If the angle p = 1″,		1 parsec (pc)
p = 1/2″,	then the distance star-sun is	2 pc
p = 1/3″,		3 pc
etc.		

In general: if the angle p = 1/D, then the distance is D pc. However, since we cannot yet go to even the nearest star and look back to the solar system, how then is this parallax angle measured?

THE MEASUREMENT OF THE HELIO-CENTRIC PARALLAX

Imagine, the earth is in position (1), as shown in Fig. 6. We take a picture of a nearby star. There will be many other stars on the picture, of course, as well as some very distant galaxies. On this picture (1), the star appears on the left side of the spiral galaxy. A quarter of a year later, the earth has moved to position (3). We take another picture of the nearby star. In this picture (3) the star appears to the right of the elliptical galaxy.

The position of the galaxies themselves with respect to each other has not changed. The distances to them are so enormous that it makes no difference whatsoever from which position of the earth we look at them. The nearby star, however, is sufficiently close to observe a parallactic displacement.

Another quarter of a year later, the earth has now moved to position (2). We take still another picture. The nearby star appears in picture (2) to the right of the barred spiral galaxy.

All pictures have been taken (or would have been by a real astronomer) with a certain camera through a certain telescope, providing us with a certain "angular

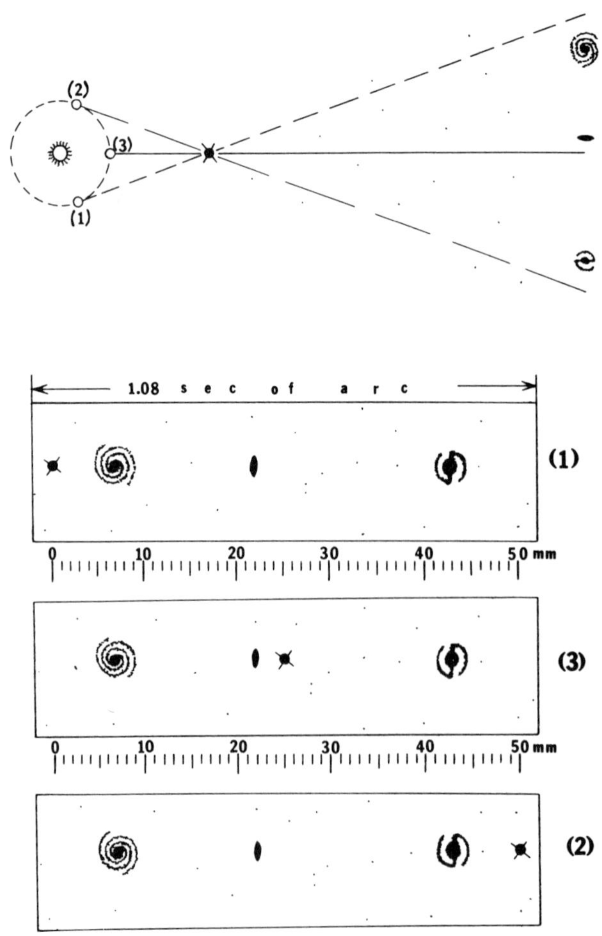

Figure 6

width'' of the picture. For example, assume that the angular width of our pictures (1), (2), and (3) is 1.08″—from their left side to their right side. The pictures themselves are 54 mm wide. Therefore, 1 mm on the pictures (linear width) corresponds to 1/50 second of arc. To measure the parallax angle, all you have to do is to measure the parallactic displacement of the nearby star between pictures (1) and (3), or between pictures (3) and (2). Remember that the parallax of a star is defined as the angle subtended by the earth's orbital **RADIUS.**

PROCEDURE

Measure this displacement between pictures (1) and (3); and (3) and (2) and answer questions 6) on p. 7–7.

C) YOUR ANGLE MEASUREMENT, USING THE POLE AND THE BUILDING, AND ITS RELATION TO THE MEASUREMENT OF THE HELIO-CENTRIC PARALLAX

Look at Fig. 7 below! Let the pole be the nearby star! The building then corresponds to the distant background of galaxies. The earth in position (1) in Fig. 5 corresponds to your position (1) in Fig. 4. Your position (2) in Fig. 4 corresponds to the earth's position (2) in Fig. 5. The sun would then be—so to speak—at the center between your positions (1) and (2). The circular arc between your positions (1) and (2) would correspond to the earth's orbital diameter or **TWICE ITS ORBITAL RADIUS.** With these relationships in mind, we will now make a **PARALLAX MEASUREMENT.**

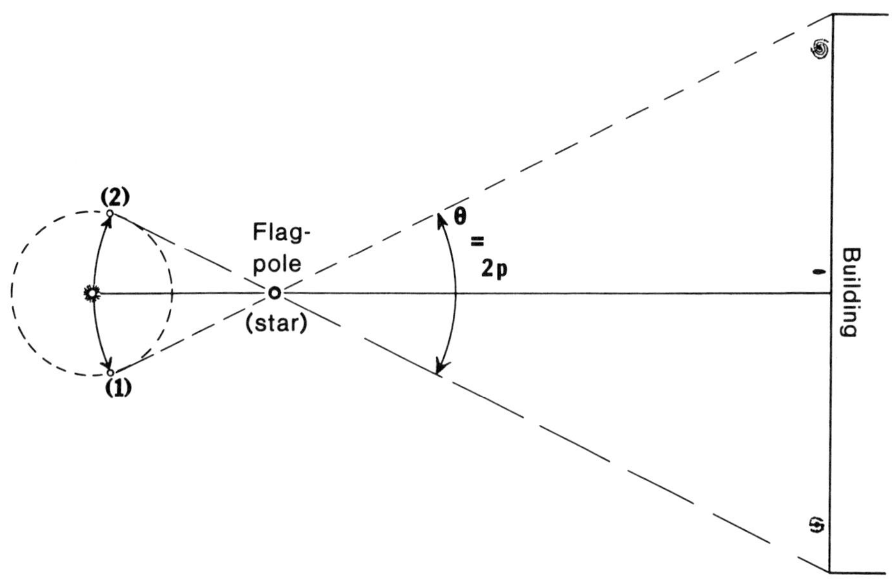

Figure 7

PROCEDURE

The pictures below were taken a certain distance from the pole(star): picture (1) from your position (1) with the pole(star) aligned with the one corner of the building. Picture (2) from your position (2) with the pole(star) aligned with the other corner of the building. The camera had an acceptance angle of 46° or 0.8 radians—the angular width of the pictures. The pictures themselves are 80 mm wide. 1 mm on the pictures therefore corresponds to how many degrees and radian? Answer question 7) on p. 7–7. Now measure the parallactic displacement of the pole(star) between positions (1) and (2). Answer questions 8) on p. 7–8.

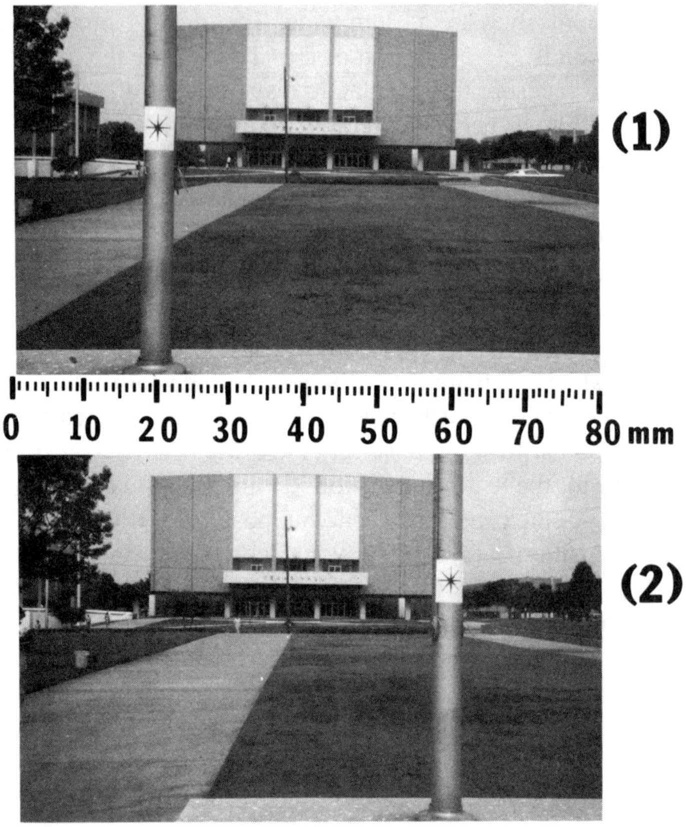

Figure 8

What is this angle? Look back to Fig. 4: It is the angle $\theta_{(1)-(2)}$ or θ. We used this angle to calculate the distance R from the pole to the building. Now we want to do something else, namely the distance from your positions (1), (2), or halfway in between to the pole(star).

We have the angle. What else do we need? Look at Fig. 7! We need the length of the arc from (1) to (2). In the helio-centric parallax measurement, that arc would be 2 AU. Here it is 2.4 m. What is now the distance from the positions where the pictures were taken to the pole(star)? Answer question 9) on p. 7–8.

Unit 7.2

The H-R and M-K Diagram

OBJECTIVE

Plot a H-R diagram on page 7–17 showing the relations of the following stellar properties to each other:

a) absolute magnitude M
b) luminosity of stars L, relative to the solar luminosity $L_{\odot} = 1$
c) spectral class
d) surface temperature
e) color
f) diameter of main sequence stars
g) mass of main sequence stars
h) locus of Cepheid—variables
i) locus of Novae
j) the band of main sequence stars
k) the M-K classes of supergiants, normal giants, and white dwarfs
l) the evolutionary track of a star of 1 solar mass.

THEORY

Once the distance to stars could be measured by the method of helio-centric parallax (see Unit 7.1 and 7.4), the emission of energy per second of those stars could be calculated, since the energy received by an observer falls off inversely proportional to the square of the distance between source and observer. The energy emitted per second is called the **LUMINOSITY (L)** of a star and is expressed in joule/sec, e.g., the luminosity of the Sun is about 4×10^{26} joule/sec. The luminosity of stars can also be expressed by the **ABSOLUTE MAGNITUDE (M)** scale or in stellar luminosities relative to the luminosity of the Sun. Note that the luminosity of a star is an intrinsic property of the star and independent of any observer.

Stellar spectra were also available for these stars. They showed many differences, but could finally be classified and ordered in a certain sequence. It turned out that this sequence is fundamentally a temperature sequence, plotted in Fig. 1 below. The hottest stars were designated to have spectral class 0, the coolest stars to have spectral class M. Each spectral class was subdivided into 10 units, e.g. 00 to 09, B0 to B9, A0 to A9, etc. The whole sequence O, B, A, F, G, K, M can easily be remembered by

Oh, Be A Fine Girl, Kiss Me!

Classes of Stars

Hertzsprung and Russel plotted the absolute magnitude M of stars versus their spectral class, i.e. their temperature. The vast majority of stars was found to occupy a relatively narrow band diagonally across the diagram, the so-called **MAIN-SEQUENCE.**

The luminosity of stars, however, does not only depend on their surface temperature but also on their surface area, i.e. on their diameter or radius since the area of a sphere is $4\pi r^2$.

For example, a star A with a surface area 10 times larger than that of a star B, will emit 10 times as much light as star B, provided both stars have the same

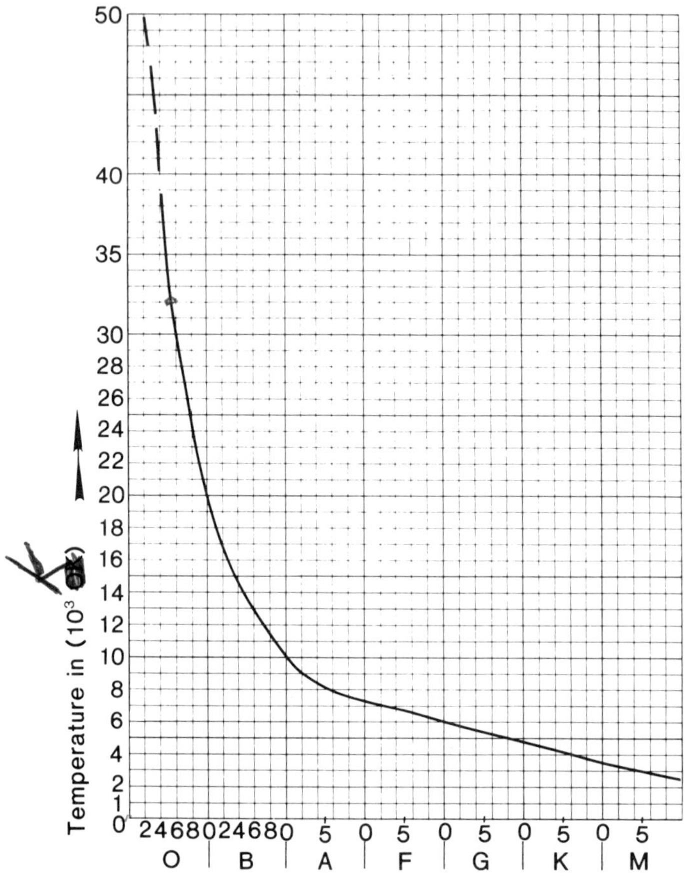

Figure 1. Temperature vs Spectral Class

surface temperature. If star B were a main sequence star, then star A would lie above the main sequence. It would have the same spectral class as star B, but its luminosity would be larger than star B's, or its absolute magnitude number M would be smaller than star B's.

Morgan and Keenan introduced another vertical scale to the H-R diagram which relates the diameter of stars to their luminosity and spectral class. Stars are then classified according to their diameters into the M-K classes:

I Supergiants
II Luminous Giants
III Normal Giants
IV Sub-Giants
V Main-Sequence Stars
VI White Dwarfs

Some astronomers sub-classify the Supergiants into: most luminous to least luminous supergiants, but that is unnecessary for our purposes.

The basic H-R diagram shows the absolute magnitude M or the luminosity of stars L as a function of their spectral class. In its original form it was restricted to those stars only, whose distance could be measured, i.e. to all stars within a sphere (surrounding the sun) of radius 100 parsec (pc) or about 300 Ly. This distance of about 100 pc was the limit for the method of helio-centric parallax. No other method to determine stellar distances existed at that time. As a matter of fact, the H-R diagram itself was **THE** next step in the hierarchy of distance determination. For the place of the H-R diagram in this hierarchy see Unit 7.4.

In the diagram you will plot in this unit, however, we will go beyond this limit and include in the diagram information about stars much farther from us than a mere 300 Ly.

There are also many additional properties of stars which can be plotted conveniently in the H-R diagram. The result will be a diagram which will allow you to "read" from it almost all stellar properties we know. Among these properties are:

a) The color of stars as seen by the human eye. It is a function of the star's surface temperature. In Unit 8.2 you will see how color perception is related to temperature. For this project you should use the following data:

STARS OF SPECTRAL CLASS	LOOK:
O	bluish-white
B	bluish-white
A	white
F	yellowish-white
G	yellowish
K	orange
M	red

b) The diameter of main sequence stars (and no others) can easily be related to their surface temperature or spectral class. Diameter versus spectral class is plotted in Fig. 2.

c) The mass of main sequence stars (and no others) also can be plotted in the diagram. This relation is usually given as "mass-luminosity relation," but is plotted here as the mass-spectral class relation in Fig. 3.

d) Stellar densities, interior temperatures, spectral lines, types of nuclear reactions in stellar cores, life times of stars, etc.

   So far, we have mentioned only properties of more or less "normal" stars, which are relatively stable and do not change any of their properties while we look at them. There are other stars which are variable and these stars happen to have certain characteristics which allow us to give them a locus in our H-R diagram. Among these objects are:

e) Variable stars like the Cepheid variables, RR-Lyrae stars, etc. and eruptive stars like novae.

   Assuming we have a finished H-R diagram to work with, containing information about many different stars, we certainly would then ask the question: **ARE** there simply that many different types of stars about or does the H-R diagram possibly indicate some hidden fact? Indeed it does:

   A star is a star is a star, but not forever. It changes. The H-R diagram therefore represents stars at different stages during their lifetime. Some "stars" are not even stars yet, but merely "Proto-stars" in the process of becoming a star, others are main sequence stars and enjoy the long years of maturity, others have already evolved away from the main sequence and are now giants, others still are almost dead and are white dwarfs. The history of a star depends very

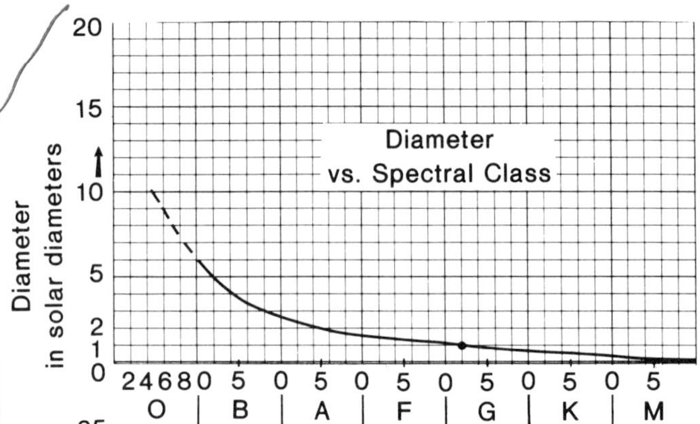

Figure 2. Diameter vs Spectral Class

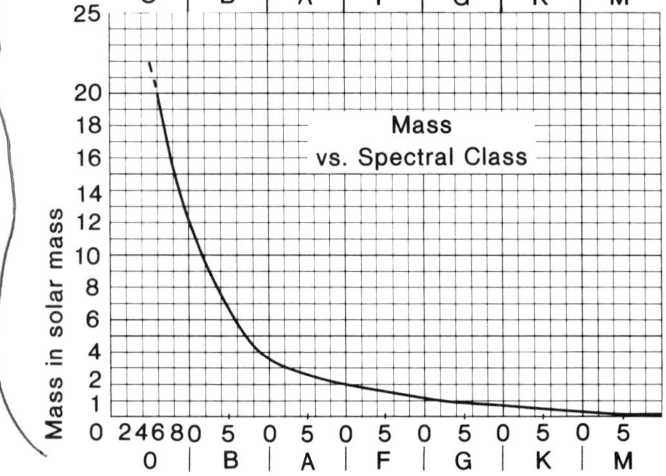

Figure 3. Mass vs. Spectral Class

much on the amount of mass it contains, and we can describe the life history of stars—as far as it is known—by plotting:

f) The evolutionary tracks of stars with certain masses.

APPARATUS

The H-R skeleton diagram at the end of this unit on page 7–17.
Pencil, colored pens, ruler.

PROCEDURE

Look at the skeleton diagram, note the spectral classes plotted as the abscissa, the absolute magnitude M plotted as the ordinate at both sides of the diagram, to make your plotting easier. The Sun$_\odot$ is already plotted with its spectral class G2 and absolute magnitude M = 5. This is a rounded-off value from 4.85. For our purposes this difference is negligible, but the plotted value of 5 will make later calculations very much easier.

As a second ordinate you will find the luminosity scale L to the left of the M scale. The luminosity of the Sun is indicated with $L_\odot = 1$.

Below the spectral class scale is the temperature scale with the solar surface temperature of 5.8 [10^3 °K] already indicated.

The following lower abscissa scales are for the color of stars, the diameter and the mass of main sequence stars in terms of solar diameter or mass. The Sun's color, diameter, and mass are already indicated. You may use the other, un-named scales either as specified by your instructor or to your heart's delight.

Before you start plotting the diagram, let's make sure that you understand its uses when finished, i.e. how to read it. Consider the Sun: As read from the diagram, the Sun is a star of spectral class G2 with absolute magnitude M = 5, corresponding to a relative luminosity $L_\odot = 1$, with surface temperature 5,800 °K. It looks yellowish, and has 1 solar diameter and 1 solar mass (of course).

You will later read the diagram in much the same way: Select a star whose properties you want to know. Read its magnitude and luminosity to the left of the star's position in the diagram, its spectral class, temperature, color, diameter, mass, etc. below the star's position in the diagram.

Do this project, using colored pens for different objects if you wish, in the following order:

1) Plot M versus spectral class for **ALL** main sequence stars listed in Table 3 of the appendix, including components, by using a dot not larger than this one: •
2) Enclose all these main sequence stars by an upper and a lower line, the band of main sequence stars, and label as such. You may wish to consult your textbook for a good plot of the main sequence because very many more stars were (probably) used for the plot than you used. The main sequence is therefore better defined.
3) Now plot the remaining stars of Table 3, using the same size dot: •
4) Plot **ALL** the stars of Table 4 of the appendix whose M is given, by using a small cross, not larger than: x Write in the names of all bright stars you can see throughout the year at your location in small print. It is advisable to use abbreviations.
5) Using the M-K classes of stars, also given in Tables 3 and 4 in the appendix, you can now divide the supergiants from the giants, the giants from the main sequence stars, etc. by drawing some lines, essentially from the left of the diagram horizontally across to the right side. Draw the first line so that all supergiants lie above this line and label "Supergiants" or "I". Draw two lines, one just above and the other just below all the normal giants, M-K class III, and label as such. All luminous giants are now located between the supergiants and the normal giants. All sub-giants are now located between the main sequence stars and the normal giants. To denote the locus of white dwarfs in your diagram, consult your textbook in addition to the few white dwarfs given in Table 3. You will find that most white dwarfs occupy a region between certain absolute magnitude values and between certain spectral classes. Indicate their approximate locus by an enclosing line and label as such.
6) Use the information provided in the theory part of this unit to fill in the abscissa scales for a) temperature, b) color, c) diameter, d) mass. Write the values for temperature, diameter, and mass below every 0 and 5 points on the respective spectral class scale, i.e. for O0, O5, B0, B5, A0, A5, etc. To do this, read the appropriate plots of Figs. 1, 2, and 3 and extract the necessary values.
7) Calculate the luminosities of stars in terms of solar luminosities. Remember that for every difference of 5 in the magnitude scale there is a ratio of 100 in luminosity value, e.g. a star A with M = 10 emits 100 times more light than star B with M = 15. Write in these calculated values, expressed by: digit, decimal point, digit × power of 10 for the luminosity on the short lines of the L scale beside the corresponding M values. You may want to consult Table 7 of the appendix for these calculations, also Unit 8.2.
8) Plot the locus of the Cepheid variables, both population I and II. Use Table 6 of the appendix.
9) Plot the locus of Novae, consult your textbook or information furnished by your instructor.

10) Draw an evolutionary track for a star of 1 solar mass. Consult your textbook.
11) As specified by your instructor, plot on the un-named scales:

____

____

____

12) Detach the finished diagram and hand-in not later than the Date due as printed on the diagram, p. 7–17.
13) Answer the questions on p. 7–15. Date due: ________

To help you with these questions which merely check your ability to read the H-R diagram correctly, let's try 2 examples:

a) Barnard's star: A main sequence star, reddish looking with surface temperature of about 3,000 °K corresponding to a spectral class M5. Its absolute magnitude is M = 13.2, it is therefore a rather very feeble emitter with a relative luminosity of only 1/1600 that of the sun. As a main sequence star it belongs to the extremely numerous so-called "red dwarfs," lying on the far lower band of the main sequence. It has a diameter about 1/2 that of the sun, and its mass is about 40% the sun's mass. Being such a feeble emitter with such a low temperature, its lifetime as a main sequence star will be very long, much longer than the sun's. Why? The energy radiated into space by stars is produced by the conversion of matter into energy. Emitting a very low amount of energy means a very small "loss" of mass and the star can go on shining for billions of years.
b) Rigel: A bluish-white supergiant, relatively rare in our galaxy, with spectral class B8 and surface temperature of about 12,000 °K. Having an absolute magnitude of about M = −7, it radiates energy into space about 63,000 as much as the sun does. The H-R diagram does not allow us to read its diameter nor its mass, not being a main sequence star. Its diameter is obviously larger than that of a main sequence star of equal temperature, otherwise it could not radiate so much more. So it lies above the band of the main sequence.

   If you had plotted an evolutionary track of a star like Rigel, it would have started with a massive, bluish-white main sequence star and would possibly have ended with a black hole. But where on the H-R diagram would be the loci of black holes?

Unit 7.2 Name ____________________

Date due ____________ Instructor ____________________ Section ____________ Seat ______

Study your H-R diagram and answer the questions for the listed stars, i.e. fill-in the empty spaces in the table, as specified by your instructor.

Star	sp. class	temp. in (10^3 °K)	color	M	L in $L_\odot$	diameter in ⊙ dia.	mass in ⊙ mass	M-K class	
1) Spica									
2) Deneb									
3) Vega									
4) Sirius A									
5) Sirius B									
6) Canopus									
7) Capella									
8) α-Centauri A									
9) α-Centauri B									
10) Aldebaran									
11) Betelgeuze									
12) Wolf 359									
13)									
14)									
15)									
16)									

Date due: ______________ Name: ______________________________

Section: ______________ Seat: __________

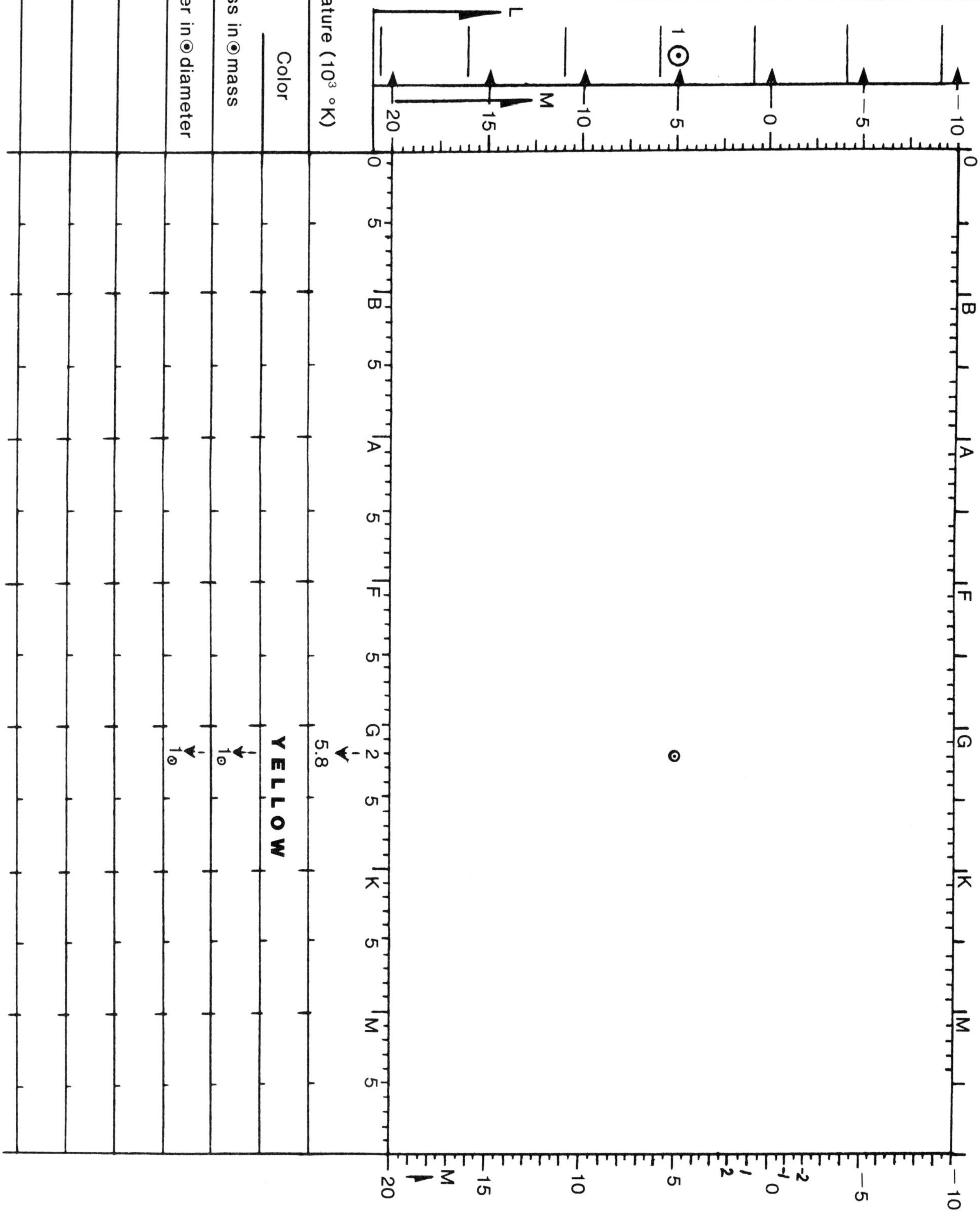

Figure 4

Unit 7.3

Diameters and Luminosities of Stars, a Model

OBJECTIVE

To draw models of stellar sizes and luminosities to scale, in order to gain a better understanding of their enormous range.

INTRODUCTION

When you read that Betelgeuze in Orion has a diameter of about 5×10^8 or 500,000,000 km and emits about 16,000 times more energy into space than our Sun does, what do these figures mean to you? It is difficult for our mind to comprehend very large (as well as very small) numbers. If a pilot of a jet-liner tells you that you are now flying at an altitude of 37,000 feet, that statement means nothing to you! You cannot possibly imagine anything walking around and counting 37,000 feet distance; it could not even count that much, probably, or surely would loose track of it. However, if we convert these 37,000 feet into miles, we come up with about 7 mi distance. And that means something to us; it is part of our everyday experience. So, to comprehend the vast scale of the universe, we must make ourselves some models to the appropriate scale.

A) DIAMETERS

In Table 1 below are listed some interesting stars and their sizes. We will compare them to the Sun and other members of the solar system; their dimensions are in rounded off figures:

Sun: diameter = 1,500,000 km, radius = 7.5×10^5 km
Earth: diameter = 13,000 km, radius = 6.5×10^3 km
Moon's orbit around the earth: 400,000 km orbital radius = 4×10^5 km
Earth's orbit around the Sun: 150,000,000 km orbital radius = 1.5×10^8 km = 1 AU
Mercury's orbit around the Sun: 0.4 AU radius
Mars' orbit around the Sun: 1.5 AU radius
Jupiter's orbit around the Sun: 5 AU radius

TABLE 1. Approximate Stellar Sizes

White Dwarfs Have a Diameter Between 0.5 and 5 Earth Diameters

Stars	Diameter in Solar Diameter	Star	Diameter in Solar Diameter
Barnard's star	0.1	Arcturus	20
Sirius A	3	Aldebaran	30
Spica	5	Betelgeuze	350 to 500 (variable)
Capella	12	Antares	600

PROCEDURE

1) Draw Model 1, the first step, on p. 7–27.
Use as the scale factor: 1 mm on paper = 7,500 km in reality
or: 1 km in reality = 1.33×10^{-4} mm on paper
The earth with a radius of 6,500 km thus is a circle with 0.9 mm radius. This circle is already drawn and labeled at the center of the page. Make your own table for the scale if you wish, using the values on the first page of this unit.

Draw to scale and name: Barnard's Star

Do *NOT* draw concentric circles	smallest white dwarf largest white dwarf the moon's orbit around the earth the Sun

2) Draw Model 2, the second step on p. 7–29.

   Use as the scale factor: 1 mm on paper = 750,000 km = 7.5×10^5 km in reality

   or: 1 km in reality = 1.33×10^{-6} mm on paper.

   The Sun with a radius of 750,000 km thus is a circle with 1 mm radius. This model Sun is already drawn and labeled at the center of the page.

   Draw to scale and name: Sirius A

Do *NOT* draw concentric circles	Spica Capella Arcturus Aldebaran Mercury's orbit around the Sun

3) Draw Model 3, the third step on p. 7–31.

   Use as the scale factor: 1 mm on paper = 7,500,000 km = 7.5×10^6 km in reality

   or: 1 km in reality = 1.33×10^{-7} mm on paper.

   Aldebaran, 30 times as large as our Sun, thus is a circle with 3 mm radius. It is already drawn and labeled at the center of the page.

   Draw to scale and name: Betelgeuze with smallest and largest radius

Do *NOT* draw concentric circles	Antares the Earth's orbit around the Sun Mars' orbit around the Sun Jupiter's orbit around the Sun

4) So much for the diameters of stars. Let us go back to Model 2. Imagine the Sun is a lighted pinhead—about the size as drawn in model 2. With the Sun's radius of 7.5×10^5 km in reality and 1 mm on paper, how far away would the nearest star be in our model? The nearest star is Proxima-Centauri. It is about 4.3 light-years away in reality. Answer this question at the bottom of p. 7–29 (Model 2).

B) LUMINOSITIES OF STARS

Luminosity is the total amount of energy emitted per second, by a star. Before we compare the luminosity of stars with the Sun's luminosity, we should first discuss the luminosity of our Sun. The Sun emits a certain amount of energy in all directions into space every second. 150,000,000 km away from it, the earth intercepts a tiny portion of this energy. The so-called **SOLAR CONSTANT** is the amount of energy intercepted by 1 cm^2 of Earth's surface per minute; the area being perpendicular to the incident sun-rays. Its value is

$$\text{Solar Constant: } 2 \left[\frac{\text{calories}}{\text{cm}^2 \text{ min}} \right]$$

Why the emphasis of the area being perpendicular to the incident rays? Imagine you are on the equator when the Sun is in the zenith at High Noon. That

would be on which date? ________ . Either on the vernal or autumnal equinox, the 21st of March or the 22nd of September. Where in the sky would the Sun appear to be at either the N or the S pole? On the horizon! In which of these places would you build a house with a solar collector to receive the maximum amount of energy?

You have certainly heard of a BTU. The calorie is the equivalent quantity in the metric system. 1 calorie (cal) is defined as that amount of heat which will raise the temperature of 1 gm of water by 1° C. (1 cm^3 of water has a mass of 1 gm). Now back to the solar constant. If we cover an area of 1 cm^2 1 cm high with water, and let it sit for 1 minute, then the temperature of the water will have risen 2° C. (It has received 2 cal).

Going further: If 1 cm^2, 150 million km from the Sun, intercepts 2 cal every minute, then the total amount of (heat)-energy emitted by the Sun is simply the total number of square cm surrounding the Sun at Earth's distance multiplied with the heat intercepted by 1 cm^2. Let us figure that out: The surface area of a sphere is $4\pi R^2$. R = distance Sun–Earth = 1.5×10^8 km = 1.5×10^{13} cm. The surface area is then $4\pi\,(1.5 \times 10^{13})^2 = 2.8 \times 10^{27}$ cm^2. Through each cm^2 flow 2 cal/min. Therefore, through this big sphere flows an energy $2.8 \times 10^{27} \times 2 \left[\frac{cm^2\ cal}{cm^2\ min}\right]$ or: The Sun emits $5.6 \times 10^{27}\left[\frac{cal}{min}\right]$ into space. This could be called the luminosity of the Sun. However, this figure is usually expressed in other units, namely joule/sec or watt. 1 cal = 4.2 joule, and 1 min = 60 sec. The luminosity of the Sun is then: $4 \times 10^{26}\left[\frac{joule}{sec} = watt\right]$. So what, pray, are 10^{26} watt? They surely would solve our energy crisis!

If one explodes 1 kgm of TNT, a chemical explosive, an energy of about 10^6 cal or about 4×10^6 joule is liberated. Now consider the energy set free by the explosion of a 1 Megaton Hydrogen-bomb. The Megaton stands for the amount of the explosive TNT which would liberate the same energy as a 1 M ton bomb. Now let us do a little bit of calculating:

1 kgm TNT = 4×10^6 joule
1 ton = 10^3 kgm TNT = 4×10^9 joule
1 Kiloton = 10^3 ton = 10^6 kgm TNT = 4×10^{12} joule
1 Megaton = 10^3 Kiloton = 10^6 ton = 10^9 kgm TNT = 4×10^{15} joule
10^3 1 Megatonbombs = 10^3 Megaton = 10^6 Kiloton = 10^9 ton = 10^{12} kgm TNT = 4×10^{18} joule
10^6 1 Megatonbombs = 10^6 Megaton = 10^9 Kiloton = 10^{12} ton = 10^{15} kgm TNT = 4×10^{21} joule
1 Bill =
10^9 1 Megatonbombs = 10^9 Megaton = 10^{12} Kiloton = 10^{15} ton = 10^{18} kgm TNT = 4×10^{24} joule

Don't give up, we are almost there!

100 Bill. =
10^{11} 1 Mton H-bombs = 10^{11} Mton = 10^{14} Kiloton = 10^{17} ton = 10^{20} kgm TNT = 4×10^{26} joule

In other words: The amount of energy the Sun pours into space, during every second it is shining, is equivalent to the explosion of 100,000,000,000 1 Megaton Hydrogen-bombs **EVERY SECOND!!!** Well, that is really beyond our imagination too. We hope that we will never have to experience the explosion of even 1 (one) lousy Hydrogen bomb. That's our sun, a rather average star—as you will see.

A question which we will not discuss in detail is: Where does this energy come from? It is produced by nuclear reactions in the solar interior. 4 Hydrogen nuclei are fused into 1 Helium nucleus. Mass is converted into energy according to Einstein's equation: Energy = mass × (speed of light)^2. If mass is converted into energy, then how much mass does the Sun "loose" or convert into energy for every second it shines? Figure it out. It is quite a bit, but then the Sun's mass itself is so enormous (2×10^{30} kgm) that it will not run out of fuel during your lifetime.

Table 2 below gives you some stars listed with their absolute Magnitude M and their luminosity in terms of solar luminosity.

Star	M	$L/L_\odot$
Sun	4.85	1
Barnard's star	13.2	0.0005 (also typical white dwarfs)
Kapteyn's star	10.8	0.004
61 Cygni	7.6	0.08
ϵ—Eridani	6.1	0.3
τ—Ceti	5.7	0.5
Sun	4.85	1.0
Sirius A	1.4	24
Arcturus	−0.3	115
Capella	−0.6	150
Aldebaran	−0.7	165
Antares	−4.5	5,500
Betelgeuze	−5.5	13,800
Rigel	−7	60,000

PROCEDURE

5) Draw Model 5 on p. 7–33.
   Use as the scale factor: 100 sq. cm of paper = luminosity of the Sun ($L_\odot$)
   or: 1 sq. cm of paper = 1/100 $L_\odot$
   or: 1 sq. mm of paper = 1/100 sq. cm = 0.0001 $L_\odot$ = 10^{-4} $L_\odot$

   The Sun's luminosity is already drawn and labeled as the large rectangle, 25 × 4 cm = 100 sq. cm, on the left side of the page.

   a) Fill the appropriate places in Table 3 on p. 7–25.
   b) Draw to scale, as rectangles, **INSIDE** the solar rectangle and name the following stars from Table 2: Barnard's Star
      Kapteyn's Star
      61 Cygni
      ϵ—Eridani
      τ—Ceti

6) Draw Model 6 on the same page as model 5, p. 7–33.
   Use as scale factor: 1 sq. cm of paper = Luminosity of the Sun $L_\odot$
   or: 25 sq. cm = 25 cm × 1 cm = 25 $L_\odot$

The Sun's luminosity is already drawn and labeled as the 1 sq. cm at the lower left **INSIDE** corner of what shall become Capella's rectangle (its left side also already drawn).

a) Fill the appropriate places in Table 3 on p. 7–25.

b) Draw to scale as rectangles, **ENCLOSING** the square of the Sun, and name the following stars from Table 2: Sirius A and Capella.

7) Draw Model 7 on the next page, p. 7–35.

Use as scale factor: 1 sq. mm of paper = 1 $L_\odot$

or: 1 sq. cm of paper = 100 sq. mm = 100 $L_\odot$

The $L_\odot$ is already drawn and labeled as the 1 sq. mm at the lower left **INSIDE** corner of the page. All the other rectangles to be drawn by you shall enclose this sq. mm.

a) Fill the appropriate places in Table 3 and answer the questions on p. 7–25.

b) Draw to scale, as rectangles, and name the following stars of Table 2:
Capella
Antares
Betelgeuze
You will not be able to draw a rectangle for Rigel. The page is not large enough. Transfer the answer from p. 7–25 to Model 7.

Unit 7.3 Name ______________________________

Date due ____________ Instructor ____________________ Section __________ Seat ____

TABLE 3

Star	$L/L_{\odot}$	Scale Factors in: Model 5 sq. cm	Model 5 sq. mm	Model 6 sq. cm	Model 7 sq. cm	Model 7 sq. mm
Sun	1	100	10,000	1	1/100	1
Barnard's Star	0.0005					
Kapteyn's Star	0.004					
61 Cygni	0.08					
ϵ—Eridani	0.3					
τ—Ceti	0.5					
Sirius A	24					
Capella	150					
Antares	5,500					
Betelgeuze	13,800					
Rigel	60,000					

There are __________ sq. cm = ______________________ sq. mm on the page.

The full page, 25 cm × 18 cm therefore would correspond to ______________________ % of Rigel's luminosity in terms of model 7 with $L_{\odot} = 1$ sq. mm. Which other star in our galaxy has about the same luminosity as Rigel: ________ . There is still another superluminous star, not in our Milky Way galaxy but in one of its companion galaxies. Which star is that, and where is it located: __________ .

Unit 7.3 Name ______________________________

Date due ____________ Instructor ______________________ Section ________ Seat ____

Model 1: First Step

Figure 1

Unit 7.3 Name ____

Date due ______________ Instructor ________________________ Section __________ Seat ____

Model 2: Second Step

In this model, the nearest star—**PROXIMA CENTAURI**—would be __________ from our model Sun.

Figure 2

Unit 7.3 Name ______________________________

Date due ____________ Instructor ____________________ Section __________ Seat ____

Model 3: Third Step

Figure 3

Unit 7.3 Name ____________________

Date due ____________ Instructor ____________________ Section ____________ Seat ______

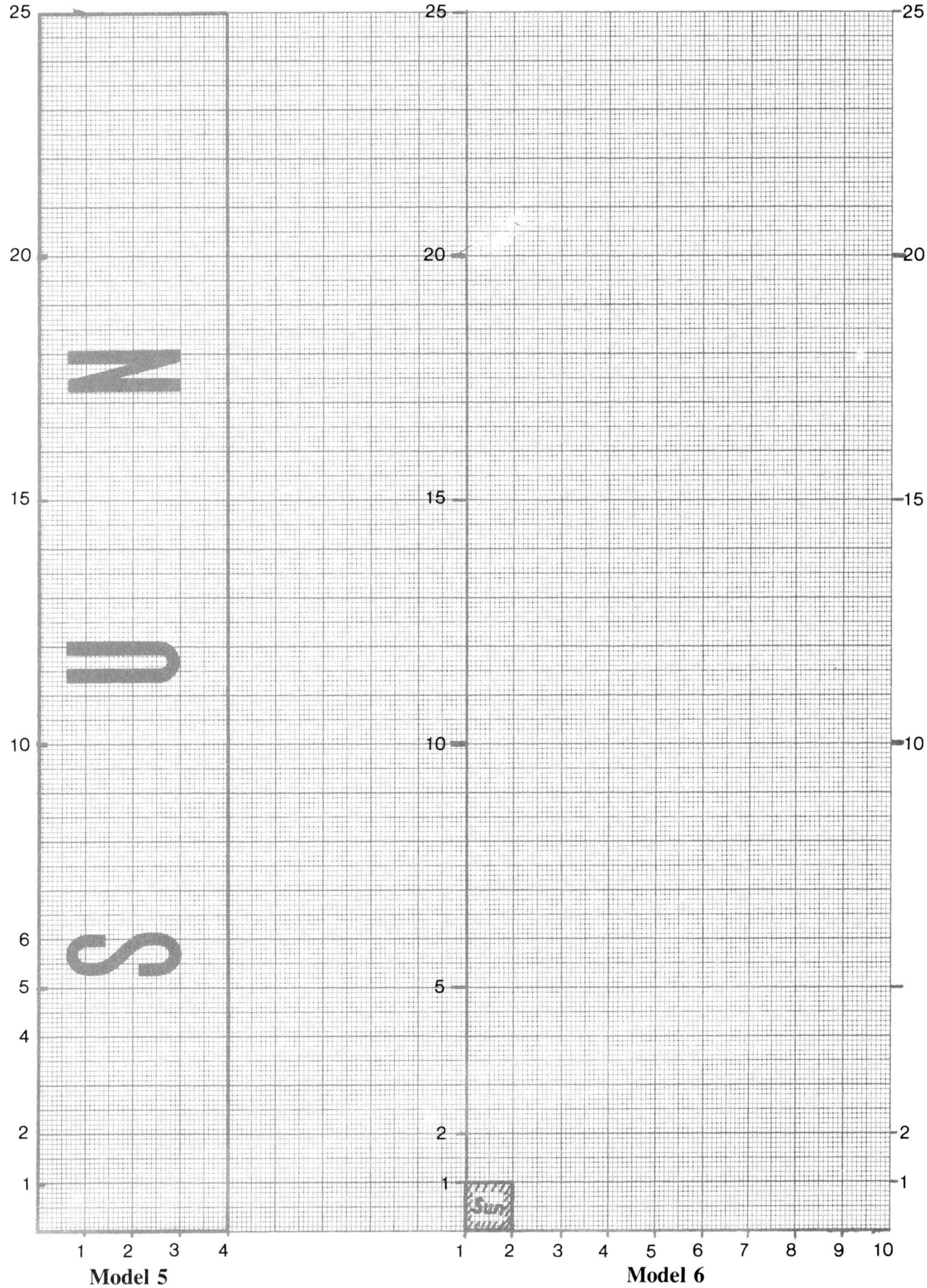

Figure 4

Unit 7.3 Name ____________________

Date due ____________ Instructor ____________________ Section ____________ Seat ______

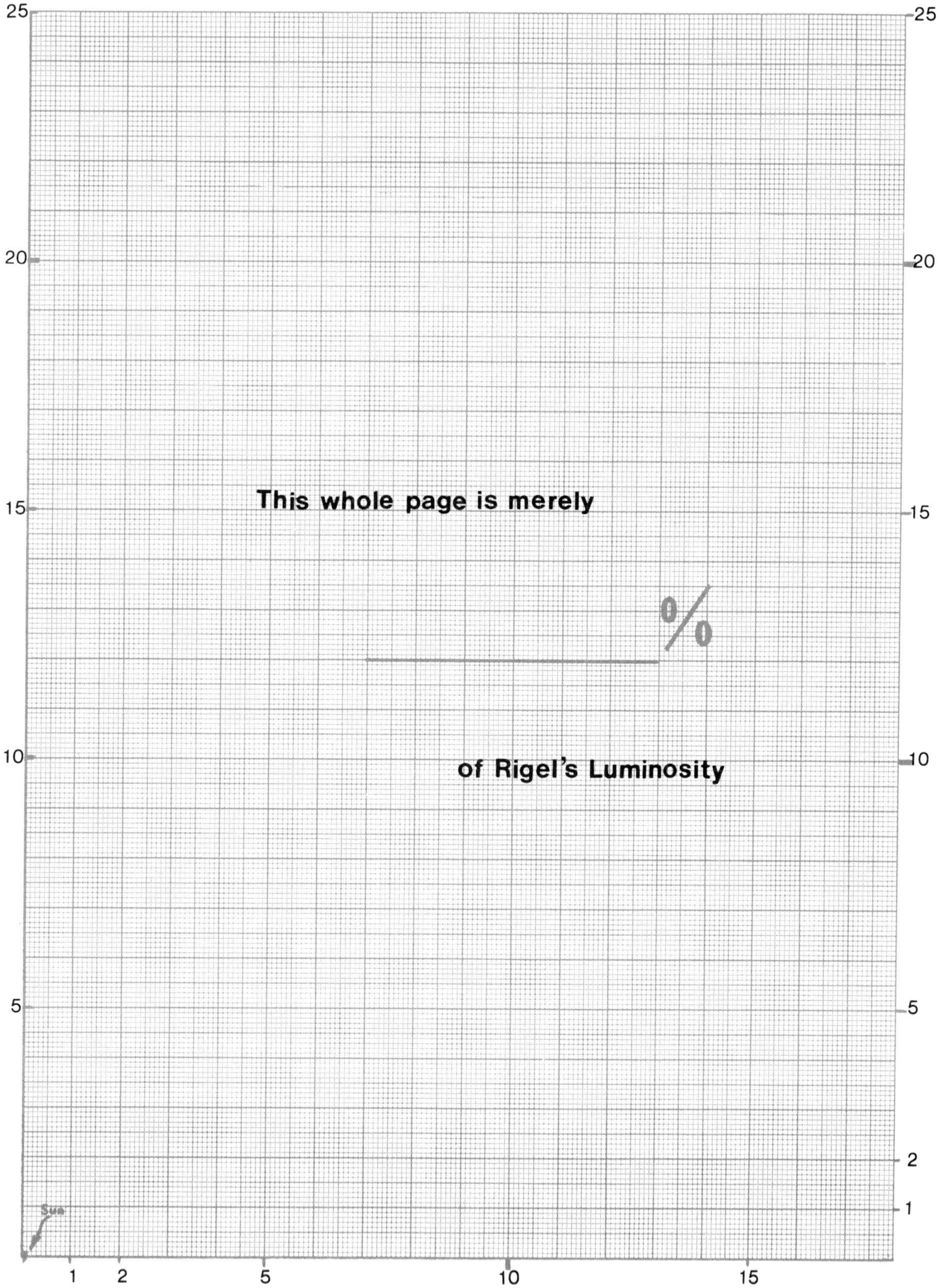

Model 7

Figure 5

Unit 7.4

The Hierarchy of Distance Measurements

OBJECTIVE

To understand how it is possible to measure the distance of objects, billions of Light-years away.

INTRODUCTION

By a ''hierarchy'' we understand a structure where various things are ordered according to rank, or where few things at the top rest on many others at the bottom and in between. The measurement of the distance to a quasar, e.g., rests on a whole series of other, lesser distance measurements. We will start with the simplest and most fundamental measurement and end up with the most involved one.

1) **THE USE OF RULERS**

A ruler is a stick, piece of string, or tape which is calibrated in certain units of length, their fractions and multiples. The distance between two points then simply corresponds to the number of units, counted off while the stick was held between these two points. We have 100% confidence in this method although there is always the question: What determines the possible accuracy of the measurement? For example: How big are the tick-marks on the ruler, what fraction of the smallest division do they allow us to read, how straight is the ruler, does it change shape when it becomes wet, does it expand or contract when the temperature changes, etc. We can, however, mostly account for these changes and take them into consideration. Usable range: Earth

2) **RULERS AND TRIGONOMETRY**

Even before the invention of trigonometry, people could measure the length of inaccessible distances by a combination of a **BASE-LINE** with geometrical considerations. For example: Measure the height of the tree! If the sun-rays just happen to make an angle of 45° with the level ground, then the angle α at the top of the tree also must be 45°, since the sum of all angles in a triangle is 180°.

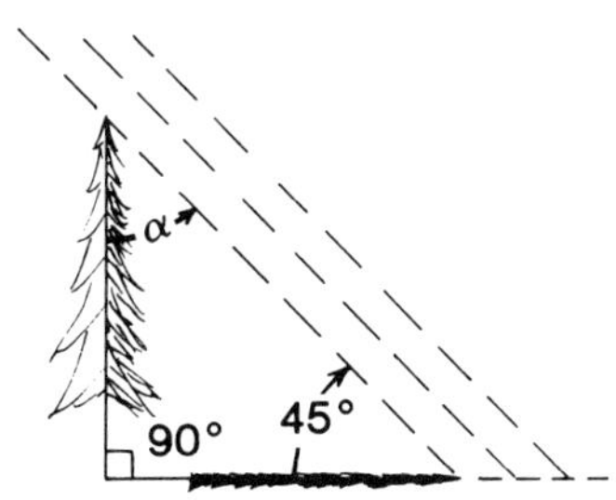

Figure 1

Therefore: If one measures the base-line—the length of the shadow—with a ruler, then the height of the tree must be equally long. To the length-measurement by ruler is now added an angle measurement. One can solve the same problem for any arbitrary angle between sun-rays and ground, if the methods of trigonometry are applied. In Fig. 2, a right triangle, the base-line is [A-B] and is mea-

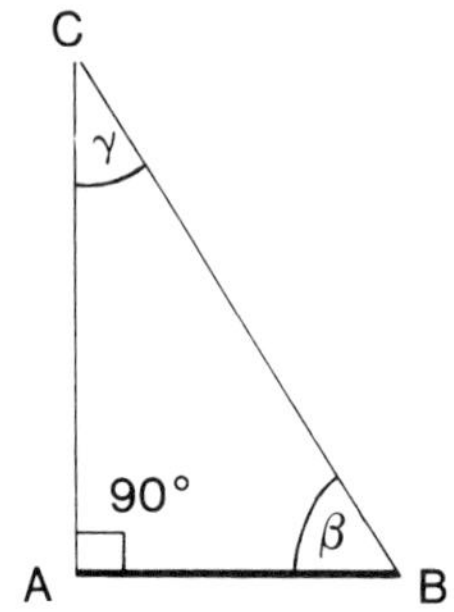

Figure 2

sured by a ruler. If the angle β is also measured, then one can use the following relations:

$$\tan\beta = \frac{A\text{-}C}{\boxed{A\text{-}B}} \qquad A\text{-}C = \boxed{A\text{-}B}\ \tan\beta \text{ or:}$$

$$\cos\beta = \frac{\boxed{A\text{-}B}}{B\text{-}C} \qquad B\text{-}C = \frac{\boxed{A\text{-}B}}{\cos\beta}$$

If the angle γ were known, one could apply the other two trigonometric functions:

$$\sin\gamma = \frac{\boxed{A\text{-}B}}{B\text{-}C} \qquad B\text{-}C = \frac{\boxed{A\text{-}B}}{\sin\gamma} \qquad \text{and } \cot\gamma = \frac{A\text{-}C}{\boxed{A\text{-}B}} \qquad A\text{-}C = \boxed{A\text{-}B}\ \cot\gamma$$

In case of oblique triangles (no angle is 90°), the law of sines and the law of cosines must be employed, instead of these simple trigonometric relationships.

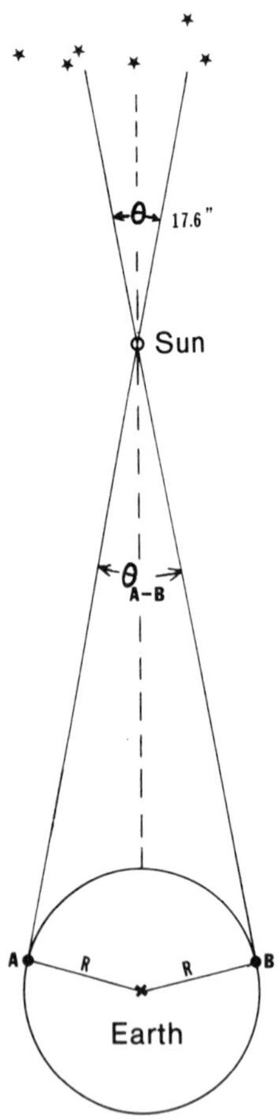

Figure 3

If any of these measured angles is very small, then the angle itself (expressed in radians) can be used instead of the sin or tan. You have encountered this problem already in Unit 7.1. Let us apply it for the next step in our hierarchy, the measurement of the distance Earth–Sun. Study Fig. 3. Two observers, one at point A, the other at point B on the Earth, simultaneously see the sun to lie in slightly different positions among the stars. This angle θ is the same as angle $\theta_{A\text{-}B}$, and was measured to be 17.6″. Observer A as well as observer B, of course, know their base-line, namely the diameter of the Earth. Applying the relation $\theta = \frac{\text{arc}}{\text{radius}}$ as you did in Unit7.1, the distance (center of Earth) to (Sun) can be calculated.

QUESTION

Answer this problem on question sheet, p. 7–47. Use for the radius of the Earth: 6,380 km.

The above method is also called the method of **GEO-CENTRIC PARALLAX**. To sum up this second step in our hierarchy: We have 100% confidence in it. It rests on unquestionable premises and simple methods, even though some very small angles must be measured for astronomical objects. With base-lines available on Earth, we can determine distances within part of our solar system.

There are other methods used in our immediate planetary neighborhood. One employs a pulsed Laser beam. It is sent from the McDonald Observatory in Texas to the retro-mirror, left by the astronauts on the lunar surface. The time is measured which a laser pulse needs to travel from the observatory to the mirror and back to the observatory. The speed of light is very accurately known. One must make some corrections for the passage of light through the earth's atmosphere (where it travels slightly slower), but the method allows a determination of the distance Earth–Moon to a few inches accuracy.

QUESTIONS

Answer questions 2) on the question sheet, p. 7–47. Assume the following: The speed of light shall be 300,000.0 km/sec. The distance to the moon varies since it revolves around the earth in an elliptical orbit.

case a) the moon shall be exactly 384,000.0 km distant from us. What is the travel time of the laser pulse (to and fro)?

case b) some time later, the moon is now 384,003.0 km away. What is now the travel time (to and fro)?

What is the minimum time interval which the clock, used for this measurement, must permit us to read?

3) **BASE-LINE AND TRIGONOMETRY:** The Helio-centric Parallax. This method had been discussed in detail in Unit 7.1 on p. 7–3. The base-line used is the distance Sun–Earth. The parallax angles are very small. The smallest angle measurable is about 1/100 of a second of arc. That means that the farthest distance which can be measured by this method is 100 pc. Since 1 pc = 3.26 LY, that corresponds to a distance of about 300 LY. There are about ten-thousand stars within a sphere with radius 300 LY, surrounding the sun. The distances to these stars therefore could be measured. **THIS IS THE LIMIT OF ANY DIRECT DISTANCE MEASUREMENT!** Let us review the situation: Our galaxy is about 150,000 LY across (how do we know that?) and contains about 200,000,000,000 stars. Our sun is located out in the spiral arms of our galaxy. Imagine a tiny (relatively, of course) sphere surrounding the sun with a radius of 300 LY. That is our backyard, so to speak; How can we climb over the fence?

QUESTIONS

Answer questions 3) on p. 7–47.

4) **THE SPECTROSCOPIC METHOD OF DISTANCE DETERMINATION**

This step in our hierarchy is of crucial importance. You have heard a little bit about it already in Unit 7.2. We start with our ten-thousand stars whose distance we know by the helio-centric parallax method. Knowing their distance and measuring the amount of light we receive from them, we can calculate their luminosity: The amount of light received is expressed by brightness, of which the apparent magnitude m is a measure. The amount of light received decreases inversely proportional to the square of the distance. The absolute Magnitude M is the brightness a star would have if it were at a reference distance of 10 pc. Or: apparent magnitude m, absolute Magnitude M, and distance D expressed in pc are related by the formula:

$$M = m + 5 - 5 \log D \qquad (7\text{-}1)$$

For the calculations you will make, you do **NOT** need to understand logarithms. Whatever you will need for the questions will be provided in this unit. For now just a brief comment, or rather a table from which you can extract some pertinent information:

D	log D	D	log D
$10 = 10^1$	1	$1 = 10^0$	0
$100 = 10^2$	2	$0.1 = 10^{-1}$	−1
$1{,}000 = 10^3$	3	$0.01 = 10^{-2}$	−2
$10{,}000 = 10^4$	4	$0.001 = 10^{-3}$	−3
$100{,}000 = 10^5$	5	etcetera	
etcetera			

The log of a number between 10 and 100 therefore ought to be between 1 and 2, right? The log of 4 is somewhere between 0 and 1?

Example: Suppose a star A appears with apparent magnitude m = 3 and is at a distance of 10 pc. What is its absolute magnitude? The log of 10 = 1.

$$\begin{aligned} M &= m + 5 - 5 \log D \\ &= 3 + 5 - 5 \times 1 = 3! \end{aligned}$$

Of course, that is the way the absolute magnitude scale is set up.

Suppose a star B appears equally bright as star A in the previous example, i.e. has m = 3, but is 10 times farther away than star A, at a distance of 100 pc. Before we use our equation (7-1), let us consider this case. Star B is 10 times farther, the light we receive would decrease by 1/square of the distance, we would receive only $1/10^2$ or 1/100th as much light. However, we **DO** receive the same amount as from star A. Therefore: Star B must **EMIT** 100 times as much light as star A, its absolute Magnitude M number must be smaller by 5. Now use equation (7-1): D = 100 pc, log D therefore = 2, m as before = 3.

$$\begin{aligned} M &= m + 5 - 5 \log D \\ &= 3 + 5 - 5 \times 2 = 8 - 10 = -2 \quad \text{Indeed!} \end{aligned}$$

So far, we have not discussed anything relating to a distance measurement yet.

If we could somehow know the absolute Magnitude of stars, then we could solve equation (7-1) for log D, since we always can measure the apparent magnitude m. Remember, we still talk about the ten-thousand stars inside our sphere with 100 pc radius. Their spectra were also obtained and classified into the spectral classes from 0 to M stars and into the M-K classes from Supergiants to white dwarfs. Let us restrict our considerations to **MAIN SEQUENCE STARS,** however, since we are after the principle of the thing only. Consider a nearby main sequence star of a certain spectral class, i.e. temperature. As a main sequence star it is located in the band of main sequence stars in the H-R diagram. To a certain spectral class therefore belongs a relatively narrow range of absolute Magnitudes, as established from our 10,000 stars. (The main sequence band you plotted in 7.2 was derived from merely 20 stars). Now imagine a star so far away that its helio-centric parallax is immeasurably small. We can still observe its spectrum, however. Therefore we can classify the star to its spectral class. We can also, by looking at the finer details of its spectrum, determine that it is a main sequence star, e.g. **THEREFORE, IT MUST HAVE A CERTAIN ABSOLUTE MAGNITUDE,** at least must have an M within a narrow range. All main sequence stars of a certain spectral class in our solar neighborhood have a certain magnitude range. We cannot imagine any reason, why a main sequence star of the same spectral class should have a different absolute Magnitude, merely because it is farther away. This is the basis for the spectroscopic method. Below, in Fig. 4, is reproduced part of an H-R diagram, showing only the band of main sequence stars, their spectral class, and their absolute Magnitude. Use this diagram to answer the questions 4) on p. 7–47.

First, however, two examples will be discussed below:

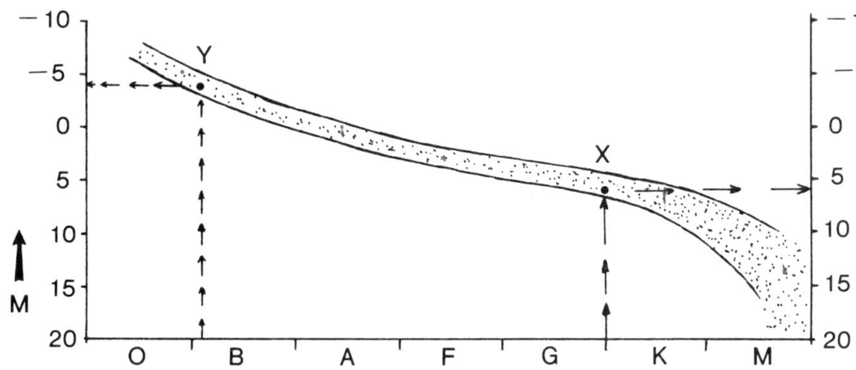

Figure 4

A star X is observed with apparent magnitude m = 11. Its spectrum shows that it is a main sequence star of spectral class K0. From the H-R diagram we can now read that it has an average absolute Magnitude of M = 6. Solving equation (7-1) for log D:

$$(7\text{-}2)\quad \log D = \frac{m - M + 5}{5} = \frac{11 - 6 + 5}{5} = \frac{10}{5} = 2 \qquad D = 10^2 \text{ pc}$$

This star would be at the very limit of our ability to measure its helio-centric parallax.

Another star Y is observed with apparent magnitude m = 6 (we receive 100 times more light from this star Y than we did from X, by the way). It is also a main sequence star but with spectral class B1. From the H-R diagram we infer that it must have an absolute Magnitude of about M = −4. What is its distance?

$$\log D = \frac{m - M + 5}{5} = \frac{6-(-4)+5}{5} = \frac{6 + 4 + 5}{5} = \frac{15}{5} = 3$$

Its distance is D = 10^3 pc = 1,000 pc = 3,260 LY.

QUESTIONS

Now answer the questions 4) on p. 7–47.

There is, of course, some uncertainty in this very method. The absolute Magnitude cannot be read exactly from the H-R diagram, rather merely a range of M values. If you can work with logarithms a little bit, can you figure out what the uncertainty in the distance determination would be if the absolute Magnitude in both cases were known with an accuracy of ±0.5.

Well, we still have a very high degree of confidence in this method. We are, however, aware of the intrinsic uncertainty in the reading of the absolute magnitude value. That cannot be helped. What is the range of this method? As far as single stars can be seen, i.e. as long as their apparent magnitude can be measured, and their spectra obtained. That is the case for the stars inside our own Milky Way galaxy and for some of the more luminous stars in our neighboring galaxies, belonging to the so-called "local cluster of galaxies". In terms of distance, the spectroscopic method is applicable up to about a million pc (a Megaparsec = Mpc) or about some million light-years. But there are many more millions of galaxies, so distant that no spectra of individual stars can be obtained. What is the next step?

5) **THE CEPHEID VARIABLE METHOD**

There are many types of variable stars, some of them tabulated in the appendix, table 6. They change their light output, therefore their absolute Magnitude M, and we detect a changing apparent magnitude m. Some variables change irregularly, some with a very definite period. For our considerations we will neglect all others except the normal Cepheids. They were named—as a class—after the star ∂ in the constellation Cepheus, a northern circumpolar constellation. They are pulsating stars, periodically expanding and contracting. Do not confuse them with "Pulsars"! They are something else entirely. Unfortunately, not a single Cepheid is within our sphere of 300 Ly radius. They are all more distant. Many of them were observed, their periods noted, but there was no way to measure their distance and therefore their M. Then, Cepheids were discovered in globular clusters. The spectroscopic method was applied to some of the more luminous member stars of the cluster and the distance calculated. The distance to the cluster, that is! The Cepheid, being a member of the cluster therefore is at about the same distance as the whole cluster. Now the observed variation of the apparent magnitude m could be related to the absolute Magnitude change in M. The result was the so-called **PERIOD-LUMINOSITY** relation. It was still later discovered that there are two types of normal Cepheid variables having different period-luminosity relationships. The first type are population I Cepheids and they are found in the spiral arms of galaxies. The second type are population II Cepheids and they are found in the spherical halo of galaxies, mostly in globular clusters. The period-luminosity relation is drawn in Fig. 5 below. The ordinate is the average absolute Magnitude, the abscissa is the period in days. One period is the time from maximum light output to minimum light output and back to maximum light output. Note that the time scale is **NOT** linear.

EXAMPLE 1

How do we use the Cepheids as distance indicators? Consider a Cepheid variable which is observed to have a period of 3 days. It is located in the spiral arms of a galaxy, belongs therefore to population I. Its apparent magnitude m changes by about ± 0.5. Its average m = 13. Now read Fig. 5: A period of 3 days corresponds to an average **ABSOLUTE MAGNITUDE M** = −2. Now we use again equation (7-2):

$$\log D = \frac{m - M + 5}{5} = \frac{13 - (-2) + 5}{5} = \frac{20}{5} = 4$$

The distance to this Cepheid therefore is $D = 10^4$ pc = 10,000 pc = 32,600 LY. In which galaxy is it located? In our own Milky Way galaxy.

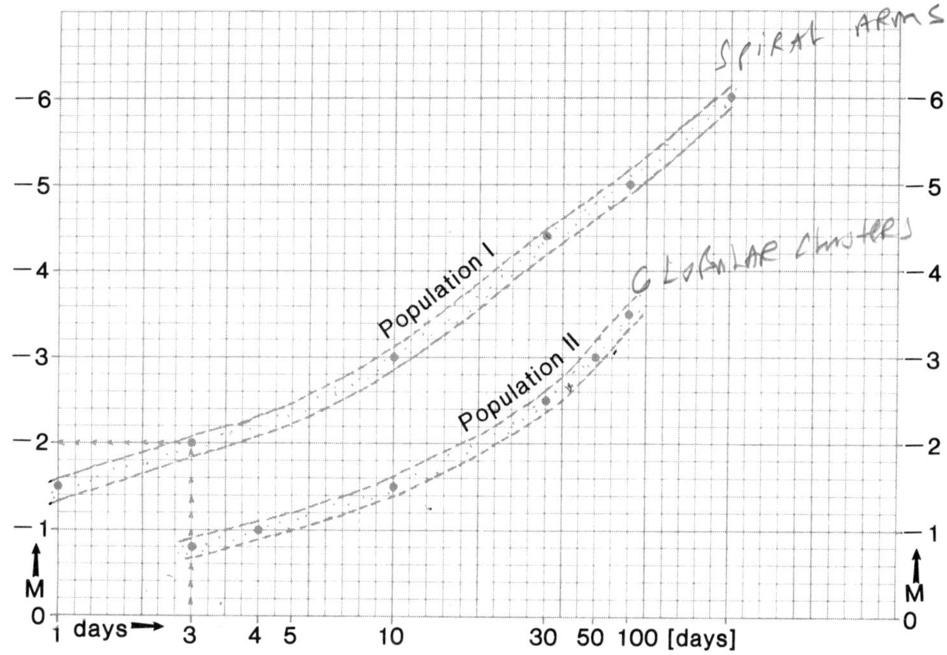

Figure 5

EXAMPLE 2

Another Cepheid variable is discovered in a globular cluster with a period of 10 days. Its apparent magnitude m is measured to be m = 23.5. How far is it? A population II Cepheid with a period of 10 days has an absolute Magnitude M = −1.5 as read from Fig. 5. Applying equation (7-2):

$$\log D = \frac{m - M + 5}{5} = \frac{23.5 - (-1.5) + 5}{5} = \frac{23.5 + 1.5 + 5}{5} = \frac{30}{5} = 6$$

The distance is 10^6 pc, more than 3 Million LY. The Cepheid is not part of our own Milky Way galaxy but belongs to another galaxy, one of the members of our **LOCAL CLUSTER** of galaxies.

QUESTIONS

Use Fig. 5 and equation (7-2) to calculate the distances to some Cepheid variables. Fill the appropriate empty spaces in Table 5, p. 7–48.

The Cepheid variable method is an extremely important and valuable step in our hierarchy of distance determinations. What is its range? Inside our own galaxy, inside our own local cluster of galaxies, and still farther—as far as it is possible to resolve distant galaxies into stars and discover Cepheids among them. Before 1924 not even the nearest spiral galaxy—the Andromeda galaxy—could be resolved into individual stars. The resolving power of a telescope depends on its **DIAMETER**—and no telescope was large enough. In 1924 the 2.5 m telescope on Mt. Wilson was commissioned, and for the first time the old suspicion could be verified that the "Spiral Nebulae" are indeed huge systems of stars. The largest telescope today (in the USA) is the Mt. Palomar 5 m mirror. Under ideal conditions we can expect to find Cepheid variables up to a distance of about 10^8 pc = 100,000,000 pc or 100 Mega-parsec (Mpc). A Mega-parsec (Mpc) will be our yardstick from now on! There are still millions of other galaxies around, so distant that they always appear as nebulous objects with no possibility to resolve them into stars and discover Cepheids—at least not yet. With a huge mirror in space or on the moon we will be able to expand the range of the Cepheid method, but that is a long way in the future. Something else was discovered which allowed us to extend our distance measurements:

In Unit 8.3 the Doppler effect is discussed. As applied to stars, some show a red shift, others a blue shift. We interpret this phenomenon as due to a relative speed of recession in case of the red shift, and a relative speed of approach for the blue shift. Stars show both kinds of Doppler shift. From the observed shift one can calculate the speed, see Unit 8.3.

Now we will look at galaxies. Our own Milky Way galaxy is a member of the local cluster, containing about 2 dozen galaxies. Within the cluster, the motion of galaxies is random. Some move with a component of their speed towards us, some away from us. Therefore we find both red and blue Doppler shifts. However, once we look at galaxies **OUTSIDE** our own cluster, we find that they all show a **RED-SHIFT.** We will call these galaxies "distant galaxies". All galaxies are apparently members of clusters, so one can also say that all clusters of galaxies move away from us (from the local cluster). Not only that. It turns out that the farther away a distant galaxy is, the **FASTER** it moves. If a galaxy A moves with a certain speed of recession, another galaxy, twice as far as A, will move with twice the speed of A.

This phenomenon is expressed in Hubble's law:

speed of recession = a constant × distance:

$$v\left[\frac{\text{km}}{\text{sec}}\right] = H \times D \tag{7-3}$$

H is Hubble's constant. Its value is still subject to refinements. At this time its value is between 30 and 70 km/sec per Mpc. For this unit we will adopt a value:

$$H = 50\left[\frac{\text{km}}{\text{sec Mpc}}\right]$$

D is the distance of a distant galaxy—as measured e.g. by the Cepheid method. Fig. 6 below shows a graph of Hubble's law. The unbroken line indicates the range of the Cepheid method.

As read from Fig. 6, a galaxy at 50 Mpc distance would have a recessional speed of 2,500 km/sec, a galaxy at 100 Mpc a speed of 5,000 km/sec, another galaxy at 200 Mpc a speed of 10,000 km/sec. This law will form the basis of the next (and last) step in our hierarchy.

6) **THE COSMOLOGICAL DOPPLER-SHIFT METHOD**

Suppose that Hubble's law is valid for all distances. Remember how it came about: Distances to galaxies were known, their red shift was measured, and their recessional speed was calculated. Result: Fig. 6 or equation (7-3). Now we will turn it around: We observe a Doppler Red Shift of a very distant galaxy. Assuming that the red shift is due to its recessional speed (there is no other explanation as of now), we can calculate its speed. Therefore: **WE CAN CALCULATE ITS DISTANCE.**

EXAMPLE

The spectrum of a very distant galaxy is obtained. At least some of its spectral lines can be identified, therefore its Doppler red shift can be measured and the

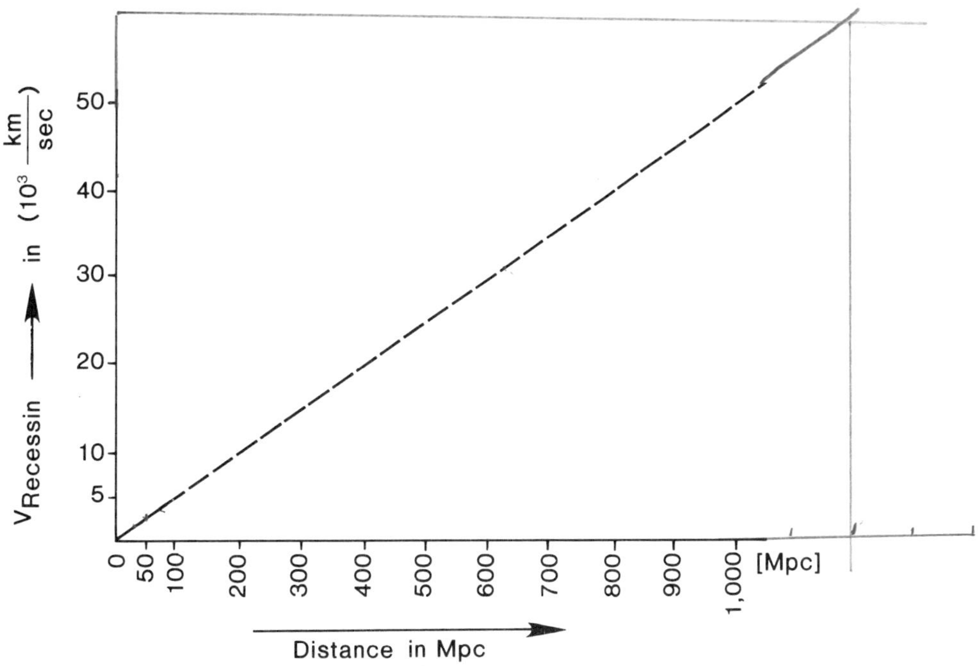

Figure 6

recessional speed calculated, e.g. v = 50,000 km/sec. Its distance can now be calculated from equation (7-3), or it can be read from Fig. 6: Going horizontally from the speed of 50,000 km/sec on the ordinate to the right until Hubble's curve is met, then going vertically down, we can read the distance: 1,000 Mpc = 1 billion pc. Now back to the very first question of this unit. How can we measure the distance to a quasar? The only information about a quasar, applicable to its distance determination, is its spectrum. From it can be calculated the cosmological Doppler shift and—if Hubble's law is indeed valid over these enormous distances and times—its distance from us.

QUESTIONS

Answer questions 6) on p. 7–48, using Fig. 6 or equation (7-3).

We have made a journey, covering distances from a millimeter to billions of light-years. We have discussed a few methods to measure these distances. The last one depended on the previous one, that one stood on still another one, etc. Is that all there is? If it were, we would be in very poor shape, indeed. There are also other methods to determine distances. They support the sequence of measurements we have discussed. To give a hint: Galaxies come in various shapes and sizes. Our own Milky Way galaxy is a giant spiral galaxy; so is the Andromeda galaxy, by the way. They also occur in clusters. One can then make an analysis of the distribution of different kinds of galaxies in clusters, and can then use the light emitted by a whole galaxy as a yardstick.

QUESTION

Answer questions 7) about other distance measurements within and beyond our galaxy on p. 7–49.

The questions:

How big is the whole universe, or
how far is the edge of the observable universe,
what is its structure, and
how old is it?

still remain to be answered. After you have finished your astronomy courses, continue to keep your eyes and ears open.

Unit 7.5

A Scale Model of the Milky Way Galaxy

OBJECTIVE

To draw a scale model of our galaxy, showing its main features in correct proportion. Consult your textbook for information.

PROCEDURE

Answer the questions on p. 7–53. These answers will form the basis of your model. Draw our Milky Way galaxy to scale on p. 7–54 with a) top view, b) edge-on view. Use imagination, draw it any way you like, as long as the model shows the following features in correct proportion:

1) nucleus
2) spiral arms and the number of their windings around the nucleus
3) location of interstellar gas and clouds
4) spherical halo with globular clusters
5) the large and the small Magellanic clouds—as satellite galaxies
6) the approximate location of the sun inside the galaxy, in both top and edge-on view
7) indicate the direction of rotation of the galaxy around its center

8

A LITTLE BIT OF ASTRO-PHYSICS

Unit 8.1

Spectral Sources

OBJECTIVE

To observe various kinds of spectra, to measure the wavelengths of some line emission spectra and to identify the chemical elements emitting these spectra.

THEORY

Spectra are classified into two large groups, a) the emission spectra and b) the absorption spectra. As the name says, emission spectra are produced by sources which **EMIT** light. The spectra are therefore **BRIGHT.** Absorption spectra are produced by sources which **ABSORB** light. These sources can be observed only if they are located in front of a bright light source. For example: The hot, solar surface (photosphere) emits a continuous spectrum, the solar atmosphere directly above the surface is cooler and absorbs certain wavelengths from the continuous spectrum. Absorption spectra are therefore **DARK.**

Each of these two groups is in turn subclassified into the following spectra:

1) continuous spectrum, consisting of a continuous range of wavelengths, its extent depending on the temperature of the emitter. It is produced by hot solids or hot, very dense gases.
2) line spectrum, consisting of discrete, individual spectral lines, characteristic of the element which either emits these lines: **LINE EMISSION** spectrum A, or which absorbs these lines: **LINE ABSORPTION** spectrum, B. A is produced by hot, mon-atomic, low density gases. B is produced by cooler, mon-atomic, low density gases in front of a continuous emission light source.
3) band spectrum, consisting of discrete, broad bands, characteristic of the chemical compound which either emits these bands: **BAND EMISSION** spectrum C, or which absorbs these bands: **BAND ABSORPTION** spectrum D. C is produced by hot, molecular, low density gases. D is produced by cooler, molecular, low density gases in front of a continuous emission light source.

To classify is a great human pastime. Nature, however, cannot always be so nicely compartmentalized as we would like. If you look at various spectra you will very often find a mixture of different classes. For example, consider a star cool enough to permit the existence of molecules and hot enough to excite some single atoms too. We can expect both lines and bands.

PROCEDURE

The laboratory consists of three parts. You may start with any of them. Refresh your memory by consulting the exhibits on the blackboard and/or the walls. Do not change the equipment, it is carefully aligned. Consult the faculty member or the lab assistant in charge for any questions you may have. If you wear glasses, keep them on.

____

____

____

____

PART I. CONTINUOUS EMISSION SOURCES A AND B

These two sources demonstrate the Stefan-Boltzmann law and Wien's displacement law. Take a seat behind one of the gratings (our spectroscopes). With your eye close to the grating, sight along the line towards the black cardboard to the side of the source slit. You will see a spectrum. Before you answer any of the following questions, look at the other source also. Mentally note their difference.

Both sources emit a continuous spectrum. Note the sequence of colors from short (blue) to long (red) wavelengths. Then look at the source **DIRECTLY,** not through the grating. Look through the slit in front of the hot Tungsten filament, and note the color (the integral color, so to speak, and not component spectral colors). Unit 8.2 discusses similar problems. Answer the questions in part I on the question sheet, p. 8–5.

PART II. DEMONSTRATION OF TWO DIFFERENT KINDS OF SPECTRA, SOURCES C AND D

Take a seat behind the spectroscopes or gratings. Look through the eyepieces of the spectroscopes or through the gratings towards the black cardboards. You will see certain kinds of spectra. Above the equipment, on the wall, are exhibits. Study them, they may help you to answer the questions in part II, p. 8–5. How are the spectra called, how were they classified, what stuff produces them?

PART III. IDENTIFICATION OF LINE EMISSION SPECTRA, SOURCES 1, 2, 3, 4

Before you sit down behind the gratings and look at the spectra, note the whole experimental set-up. In each of the tubes is a different gas, made to glow by a high electrical potential (5,000 volts between the electrodes). On each side of the glowing gas is a wavelength scale. Its divisions are the same as those on the reference sheet (p. 8–7 or 8–11) and the worksheet (p. 8–9 or 8–13). Study them and be sure you can read them before you attempt to measure the wavelengths. Note that the scales on the right side of the tube have an increasing wavelength scale to the right. The scales on the left side increase to the left. Whichever side you start with, stay with the same side for all lightsources in this part III. Use the worksheet and the reference sheet for your chosen side.

Then sit behind one of the sources. With your eye close to the grating look towards the wavelength scale and **NOT** towards the source. You will see bright spectral lines, a line emission spectrum.

If you move your eye slightly to the side, you may observe that the spectral lines also move slightly across the wavelength scale. Adjust your eye so that the lines stay at an average position. We are not interested in an extremely accurate wavelength measurement. Our purpose is to measure the spectral lines accurately enough so we can identify the element which produced them. In one of the light sources you may see a diffuse, yellow glow (**NOT** a well defined line!). Disregard it, it is undesirable but cannot be prevented. This glow is caused by the glowing glass of the tube.

1) When you are ready for the measurements, take the worksheet, p. 8–9 or 8–13, and draw in the spectral lines you see. Some source may emit a very large number of lines, some other source may be only one or two. **CONCENTRATE ON THE BRIGHTEST LINES.**
2) When you have done that, look at the reference sheet. Take your work sheet with your measured spectrum and go through the various reference spectra. Find the one which you have measured, identify the element and answer in part III on p. 8–5. **NOTE:** You may not see all the reference lines drawn for a certain element. Remember that the human eye becomes very insensitive if we look at either very short or long wavelengths, at the violet or deep red end of the visible spectrum.
3) Continue with the other light sources. Remember to stay on the same side of the sources for all successive measurements.

An interesting after-thought: Looking through the grating, you see the component colors (spectral lines) coming from this light source. Looking directly at the glowing gas, you see a quite different, overall color. Our eye is an integrating instrument, unable to see the component wavelengths. Imagine that our ear would act the same way. Would you then enjoy a concert?

Unit 8.1 Name ____________________

Date due ________ Instructor ____________ Section ________ Seat ____

REFERENCE SPECTRA **LEFT SPECTRUM**

Note: The brighter a certain spectral line appears, the thicker is the spectral line drawn in the spectra below. For purposes of identification, concentrate on the two or three brightest lines.

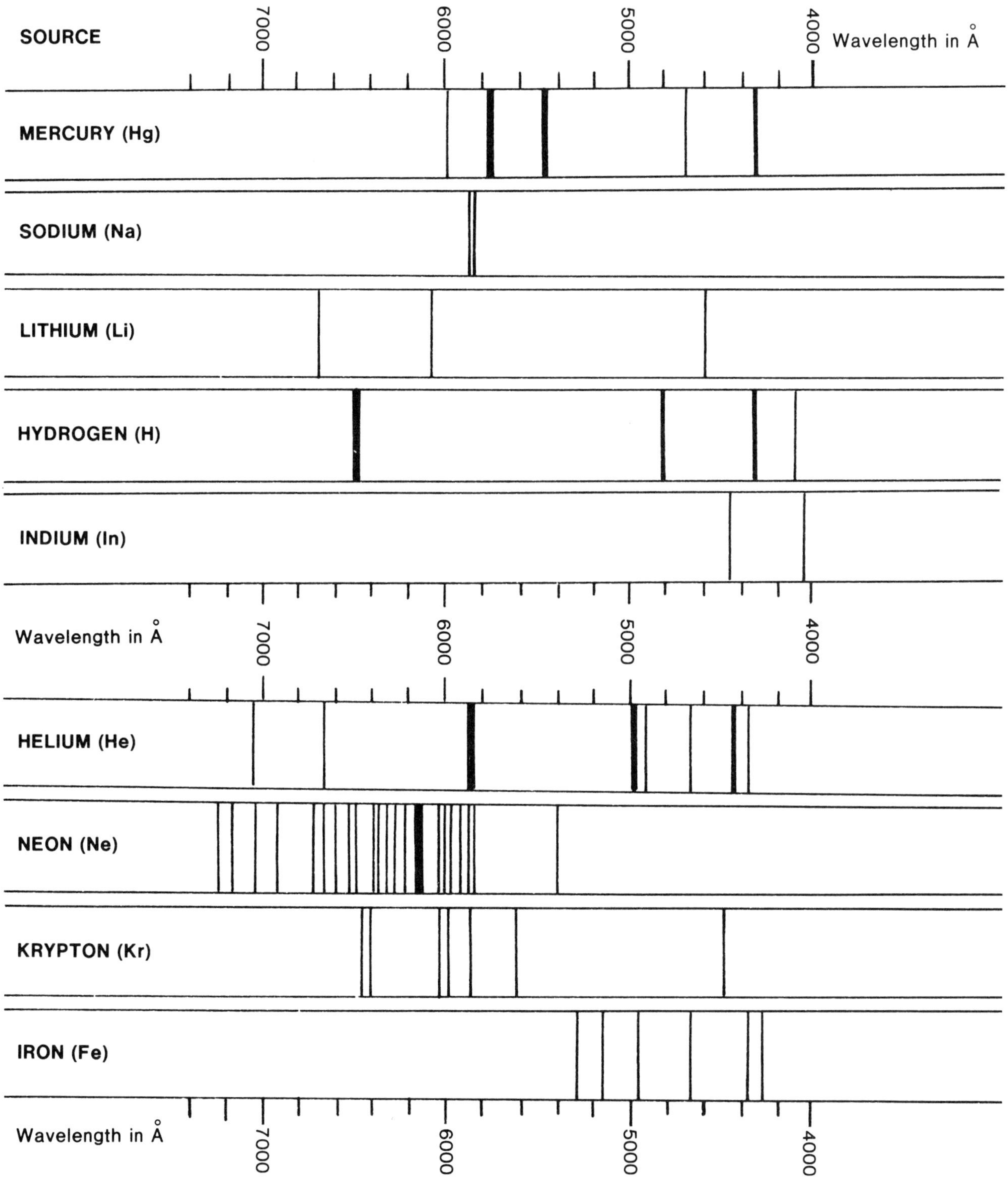

Unit 8.1 Name ______________________________

Date due ____________ Instructor ______________________ Section __________ Seat ____

WORKSHEET

LEFT SPECTRUM

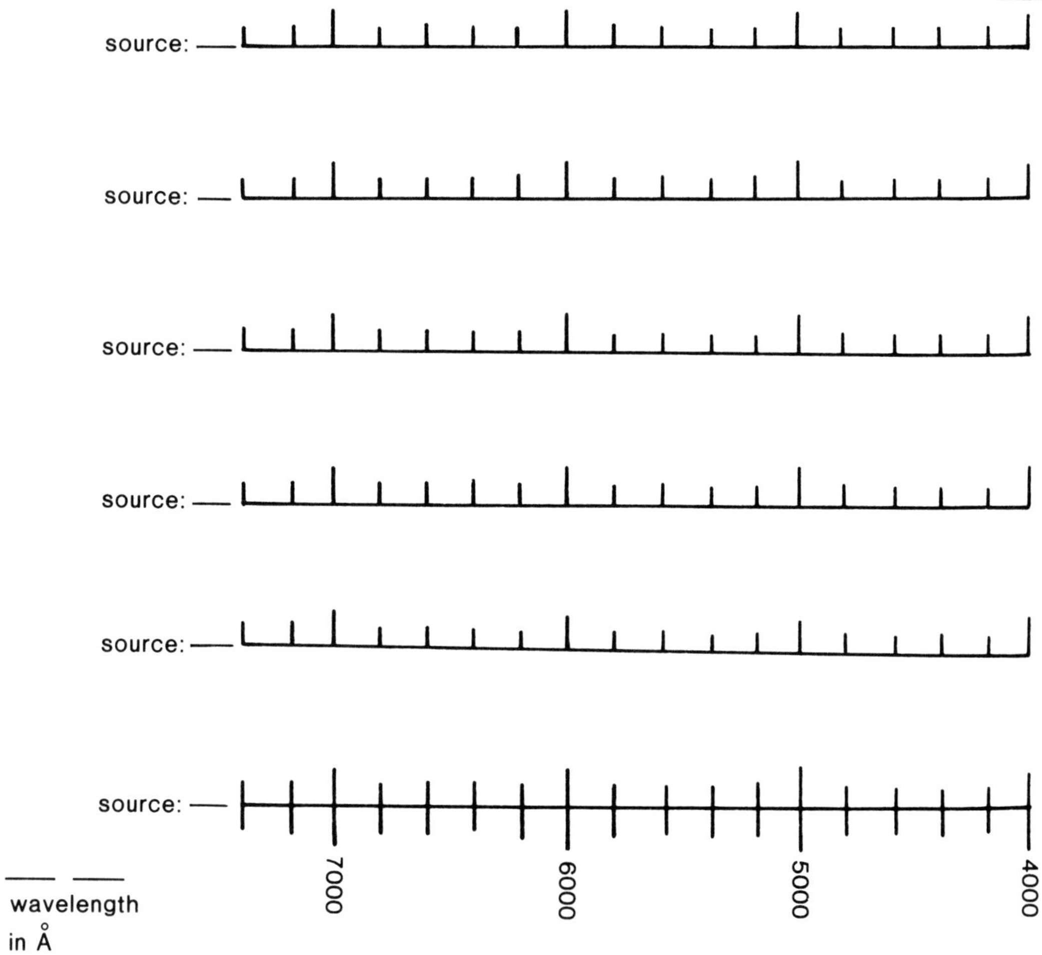

Read the wavelengths of the brightest spectral lines observed for each of the sources 1, 2, 3, and 4. Disregard very faint lines. Remember that the human eye is very insensitive for very short (violet) and very long (red) wavelengths of the visible spectrum. You may therefore not see some of the faint lines at these extreme ends of the spectrum—as indicated on the reference sheet. Draw in the brightest lines in the above scales and compare with the reference sheet. Identify the element emitting the observed spectrum.

Unit 8.1 Name ____________________

Date due __________ Instructor ____________________ Section __________ Seat ______

REFERENCE SPECTRA **RIGHT SPECTRUM**

Note: The brighter a certain spectral line appears, the thicker is the spectral line drawn in the spectra below. For purposes of identification, concentrate on the two or three brightest lines.

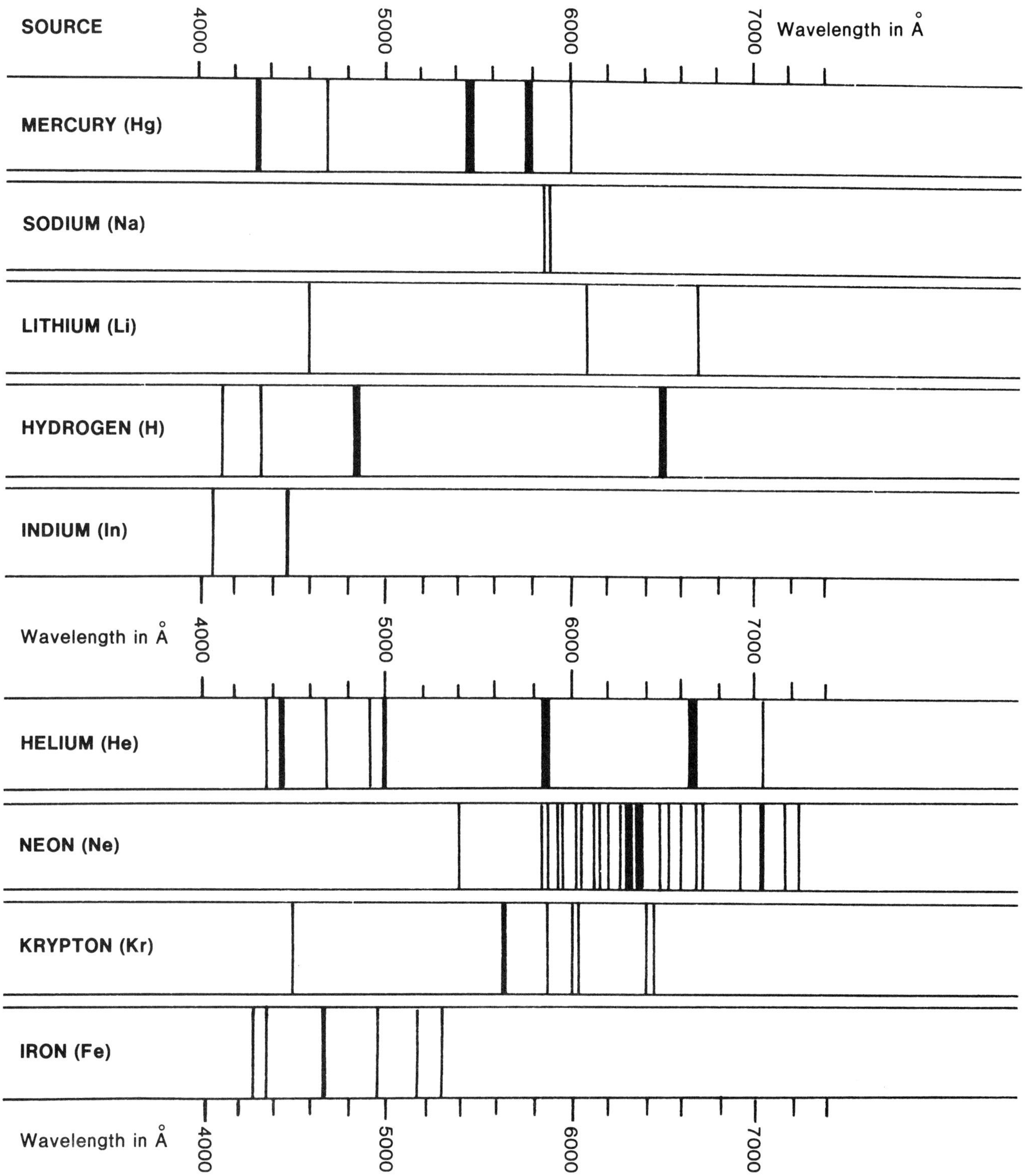

Unit 8.1 Name ______________________________

Date due ____________ Instructor ____________________ Section __________ Seat _____

WORKSHEET

RIGHT SPECTRUM

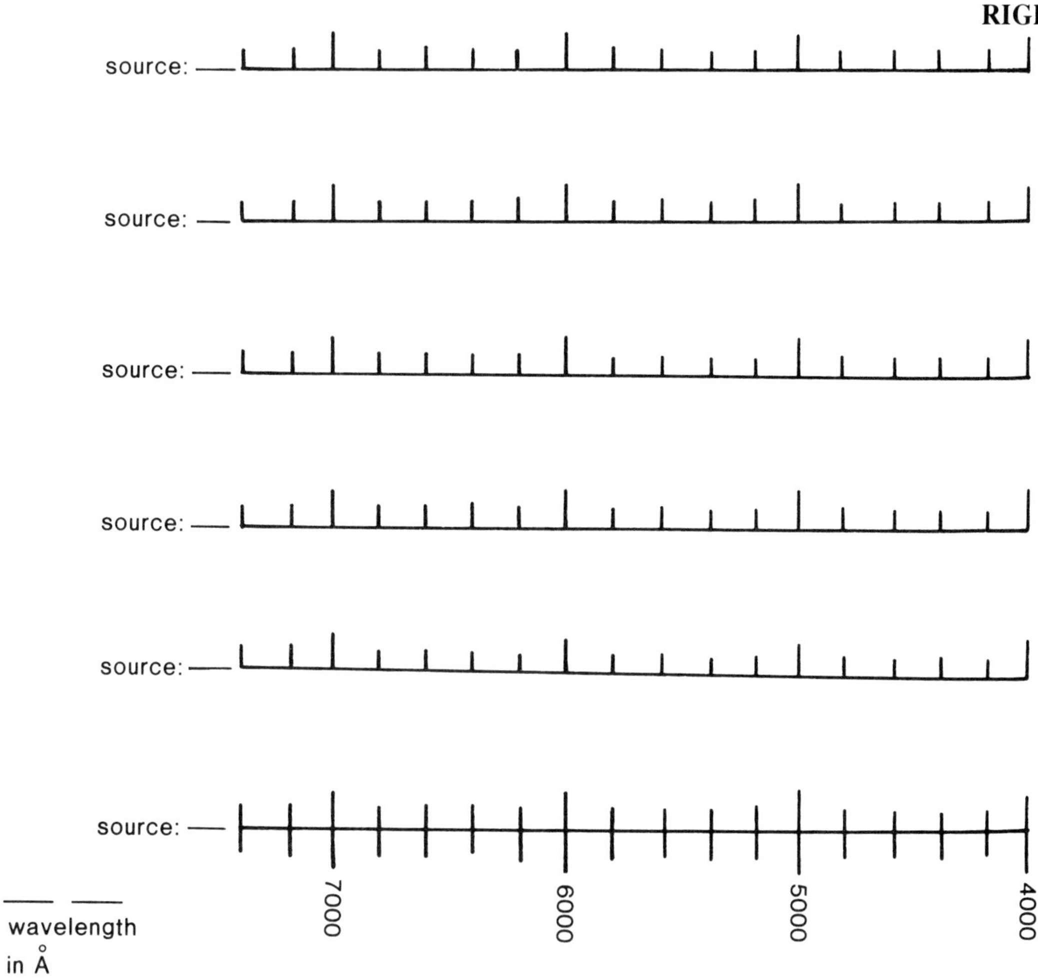

Read the wavelengths of the brightest spectral lines observed of each source 1, 2, 3, and 4. Disregard very faint lines. Remember that the human eye becomes very insensitive for very short and very long wavelengths (violet and red, respectively). You may therefore not see some of the faint lines of the reference sheet at these extreme ends of the visible spectrum. Draw in the brightest lines above and compare with the reference sheet. Identify the element emitting the observed spectrum.

Unit 8.2
The Color and Luminosity of Stars

OBJECTIVE

To understand the nature of star light, how it looks to us, and to compare the intensity of star light with that of the sun.

THEORY

Stars are hot, self-luminous bodies. The electromagnetic radiation they emit is essentially of the same kind as that emitted by hot solids, e.g. the filament of a Tungsten light bulb. This kind of radiation is treated in Physics as the so-called **"BLACK-BODY"** radiation. (A black body is a perfect absorber of energy and, after being heated up, a perfect emitter of energy as well.) Stars are very nearly black body radiators, and everything said below about black body radiation applies to radiation from stars as well.

BLACK-BODY RADIATION

The radiative energy emitted per second by a unit area of a black body depends on its absolute temperature, specifically its absolute temperature to the fourth power. We call the energy emitted per second the **"LUMINOSITY".** The equation which describes this emission is the **STEFAN-BOLTZMANN LAW.**

It states that

$$\text{Luminosity (L)} =$$

$$\frac{\text{total radiative energy emitted}}{\text{second}} = \text{Area} \cdot \sigma \cdot T^4 \left[\frac{\text{joule}}{\text{sec}} = \text{watt} \right]$$

σ is the Stefan-Boltzmann constant with the value:

$$5.67 \times 10^{-8} \left[\frac{\text{watt}}{\text{m}^2 \ {}^\circ\text{K}^4} \right]$$

The luminosity of stars is one of their most important, intrinsic properties, and you will meet this term repeatedly. Before we go on, let us discuss briefly the influence of the surface area of an emitter. Imagine two lightbulbs, a 50 Watt and a 100 Watt bulb. Both burn at the same temperature, a unit area of their surface would therefore emit the same amount of energy per second. So, what makes then a 100 watt bulb twice as powerful an emitter as the 50 watt bulb? The surface area of the 100 Watt bulb is twice as large as that of the 50 Watt bulb, its filament has twice the surface area. We have met the Stefan-Boltzmann law and the dependence of emission on the surface area in Unit 7.3. It turns out that the diameters of most stars can only be determined by the application of this law.

For all the following discussions we will neglect the surface area, or better, we will consider emitters with always the same surface area, or unit surface area. A plot of black-body radiation curves as functions of their temperatures is shown in Fig. 1 for temperatures of 2,000; 4,000; 6,000; 8,000 and 16,000 [°K].

Note that the emission is continuous. It is plotted for the visible spectral region from 4,000 to 7,600 Å, extending into the near Ultra-violet (UV) and the near Infra-red (IR). In addition to these wavelength ranges, black bodies emit a continuous spectrum of electromagnetic waves from X-rays to radio waves, depending on their temperatures. A very cool body will emit **ONLY** radio waves.

The luminosity is plotted on a logarithmic scale because the luminosity at very high temperatures is extremely much larger than at lower temperatures and therefore cannot be displayed on a linear scale. Example: Let us calculate how much more a black body A at 16,000 °K radiates than B at 2,000 °K. The temperature of A is 8 times the temperature of B. Body A therefore radiates $8 \times 8 \times 8 \times 8 =$ 4,096 times more than body B. Fig. 1 gives you a good idea of the enormous range of luminosities we will encounter with stars.

A second law about black-body radiation is **WIEN's DISPLACEMENT LAW.** It says that the higher the surface temperature of a body, the more is the peak of its radiation curve displaced towards shorter wavelengths. Looking at Fig. 1 you will see that the peak of the 2,000 °K emitter is in the IR region and off the scale, the peak of the 4,000 °K emitter is at about 7,500 Å; the peak of the 16,000 °K emitter is in the UV and also off the scale. Wien's displacement law in quantitative terms states: The wavelength at which the peak of the radiation curve occurs (λ_{max}) multiplied by the absolute temperature of the emitter is constant.

$$\lambda_{max} \cdot T = a \qquad a = 2.9 \times 10^7 [\text{Å } °\text{K}]$$

Example: The Sun's surface temperature is 5,800 °K = 5.8×10^3 °K. At what wavelength does the peak of its black-body radiation curve occur?

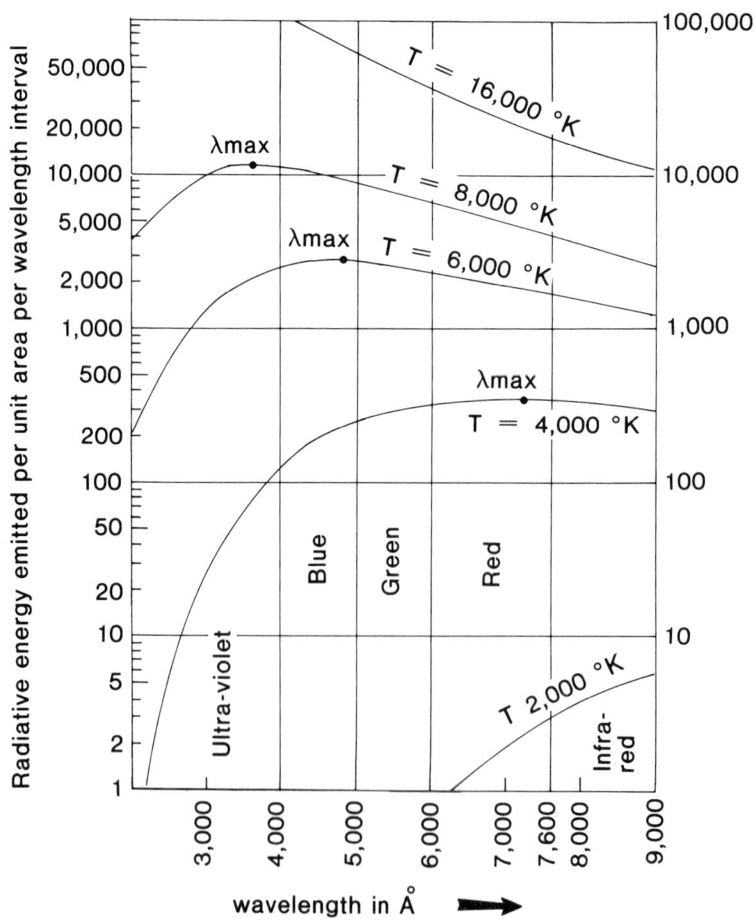

Figure 1

$$\lambda_{max} = \frac{a}{T}\left[\frac{Å\ °K}{°K}\right] = \frac{2.9 \times 10^7}{5.8 \times 10^3} = 0.5 \times 10^4 = 5{,}000\ Å$$

By measuring the wavelength of the radiation peak, λ_{max}, we can therefore easily determine the surface temperature of stars. We can either use Wien's law as above, or we can compare the surface temperature of a star with the Sun's surface temperature.

Example: Suppose a star $*$ has its $\lambda_{max,*}$ at half the Sun's $\lambda_{max\,\odot}$, namely at 2,500 Å. What is $T*$?

$$\lambda_{max,\odot} \cdot T_{\odot} = \lambda_{max,*} \cdot T_{*}.$$

Solving for T_*:

$$T_* = \frac{5{,}000 \cdot 5{,}800}{2{,}500} = 2\ T_{\odot} = 11{,}600\ [°K]$$

THE COLOR OF BLACK-BODY EMITTERS

Imagine you are heating up a piece of iron in a dark environment. Below a certain temperature, the iron does not emit any visible light at all. It still radiates. You could feel its infrared emission if you hold your hand close to it. Heating it up more, it will start to glow when your light-sensitive receptors in the retina become stimulated. Heating it up more and more, what will you see? Most people would answer: First it glows with a dull red color, then it will glow a brighter red, then orange, then yellowish, finally white and, if we could heat it up enough, bluish-white. Although these answers are essentially correct, they leave out one vital element: Before you will see it glowing reddish, it glows grey. This is not easy to observe. Your eyes must be dark adapted and no stray light should be around. Why is that so? There are two different light sensitive receptors in the human eye. The color sensitive cones and the merely light sensitive rods. The rods are much more sensitive to light than are the cones. We use them for night vision, when the general level of illumination is very low. The cones need much more light to become stimulated, hence:

At night all cats are grey! We do not see their color. If you now shine a light on the cat, the reflected light probably is of a high enough level to stimulate the cones. You then see its colors. During daylight we see with the cones. They are situated on the retina in an area called the fovea centralis (or the yellow spot) with maximum concentration (highest number of cones per square inch). Whenever we look at something in daylight, the eye adjusts itself so that the image of the object appears in this yellow spot. We see the object with most detail, with the highest resolution. This process is unconscious, the eye does it automatically. The rods, on the other hand, are situated around this yellow spot and are intermixed with cones. At an angle of about 20° from this yellow spot is the area where the rods appear with highest concentration. If we want to see something very faint, then we should not look directly at this something, but should deliberately look about 20° to its side. We use then **"AVERTED VISION".** It is important in astronomical observations. Example: Look at the middle star of the handle of the Big Dipper directly. You will probably see a single star, Mizar. Now look deliberately to the side and its very faint companion, Alcor, may become visible. Explanation: The light from Mizar is bright enough to stimulate the cones, but the light from Alcor is not. Looking with averted vision, the more sensitive rods can now also see Alcor. Use this averted vision whenever you look at a faint object, even through telescopes. Looking at the sky, most stars appear somewhat whitish, particularly most of the faint stars, because we do not receive enough light from them to stimulate the color sensitive cones sufficiently. Now back to **COLOR VISION:** Without going into too much detail, let us assume that there are three different color sensitive mechanisms associated with the cones, **RED, GREEN,** and **BLUE** receptors. (The reality is unfortunately always more complex than we would like, so we have to simplify the processes of explanation. For example, there are **NOT** three different kinds of cones in the retina; the red, green and blue stimulation takes place in each cone. Moreover, the red, green, and blue sensitivity ranges overlap.)

We will assume that there are three wavelength ranges with corresponding receptors:

Red (R) from 6,000–7,600 Å
Green (G) from 5,000–6,000 Å
Blue (B) from 4,000–5,000 Å

These wavelength ranges are indicated both in Fig. 1 and Fig. 2. To find out what color we see, we need to determine how much each of the three receptors is stimulated. As long as we look at an object which is neither too faint nor too bright (so we are blinded), only the **RELATIVE** amounts of R, G, and B are important. For this reason, Fig. 2 shows these relative energy emissions on a linear scale. Compare Fig. 1 with Fig. 2: Choose a wavelength, e.g. 7,000 Å. The relative emission of the 2,000 °K curve in Fig. 2 is 2. In Fig. 1 it has the same value. Now consider the 4,000 °K curve in Fig. 2: it has the value of 3.7 for 7,000 Å. In Fig. 1 it has the value of 370. To reproduce Fig. 1 from Fig. 2, the 4,000 °K curve would have to be multiplied by 100. Consider the 6,000 °K curve in Fig. 2. It has a value of 1.9. Its value in Fig. 1 is 1,900. Curve 6,000 °K in Fig. 2 must be multiplied by 1,000 to get the curve in Fig. 1.

The multiplication factors for the other curves are: 1,000 for the 8,000 °K curve, and 10,000 for the 16,000 °K curve. For the evaluation of the color of a light source we may neglect the absolute value of Fig. 1 and be satisfied with the relative values of Fig. 2.

HOW TO EVALUATE THE COLOR OF A LIGHT SOURCE

Consider a black body at temperature 2,000 °K and look at the 2,000 °K curve of Fig. 2. You see that the red sensitive receptor is stimulated very much more than the green receptor. The blue one is stimulated hardly at all. Result: We see red.

Consider the 8,000 °K curve of Fig. 2. Here the blue receptor is stimulated the most, although both the green and the red receptors are also stimulated. Imagine a horizontal line drawn at the value of 4 on the ordinate, and imagine further that this line were representing an imaginary emitter. It would stimulate all three receptors, R, G, and B equally. We would see **WHITE,** whenever R, G, and B are stimulated about equally.

Now back to the 8,000 °K curve. On top of this **"WHITE BACKGROUND"** sits a very high amount of blue, a smaller amount of green and a still smaller amount of red. The result: We see a **BLUISH**-White color. Now look at Fig. 1 again. Will we ever see a red star? a green star? a blue star?

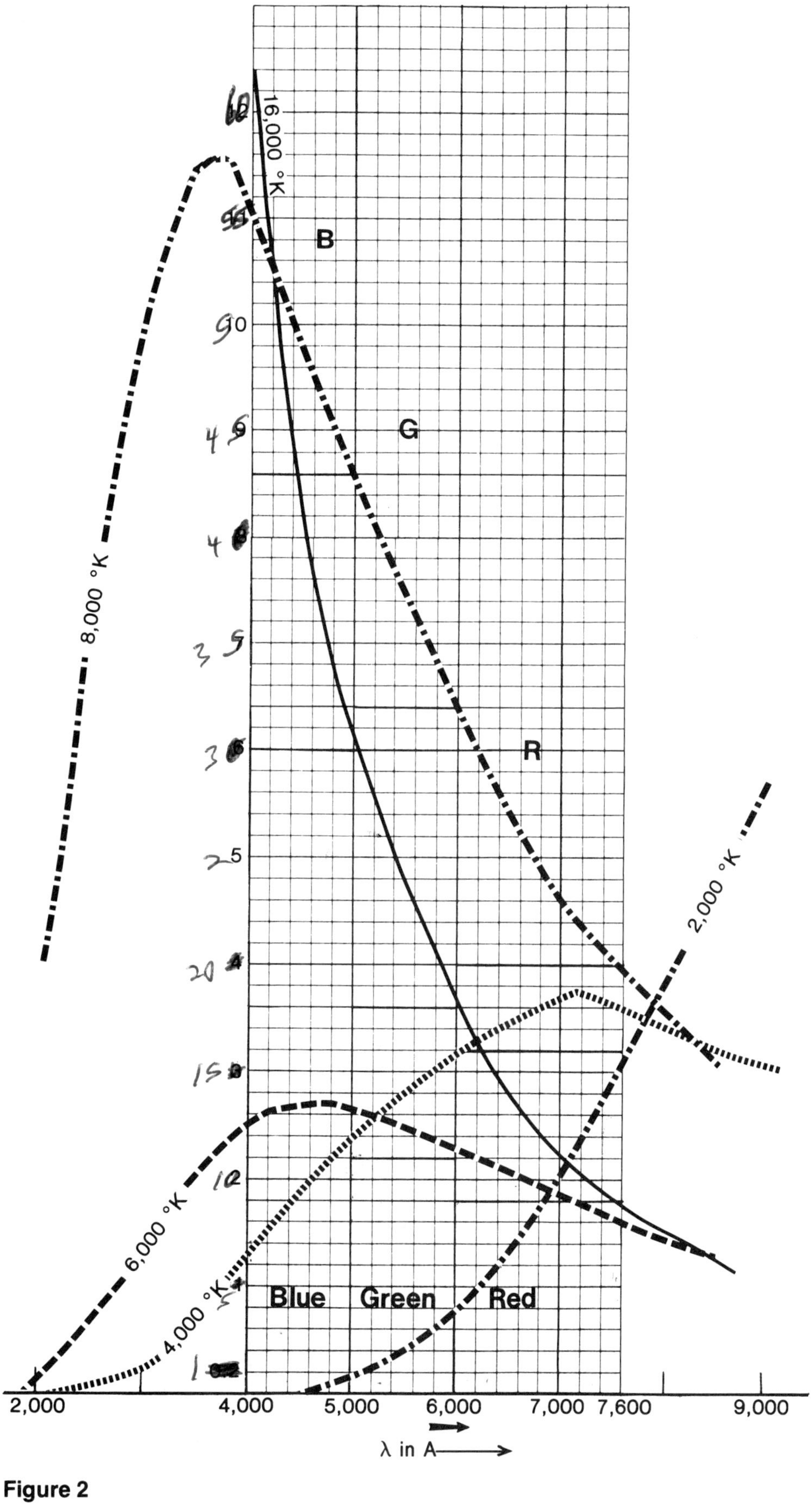

16,000 °K
8,000 °K
6,000 °K
4,000 °K
2,000 °K
B
G
R
Blue
Green
Red
2,000
4,000
5,000
6,000
7,000
7,600
9,000
λ in A

Figure 2

We certainly could see a red star: Essentially only the R receptors are stimulated. We could **NOT** see a green star: To see green, essentially only the green receptor would have to be stimulated. That is not the case for any of the black body radiation curves. We could **NOT** see a blue star: All black body curves have the long wavelength ''Tail''. However much blue light there is, there is always also quite a bit of green and red, i.e. a white background. Even if the temperature were 100,000 °K, the white background is still there and we would always see bluish-white.

APPARATUS

Pencil, paper, and possibly a calculator.

PROCEDURE

A) THE COLOR OF STARS. Use question sheet on p. 8–23.

1) Determine the receptor stimulation for each of the radiation curves in Fig. 2 by counting the number of squares below each of the radiation curves for R, G, and B: First, count the number of full squares below each curve, then the number of partly enclosed squares, matching parts with each other so that they correspond to full squares. Write the total number of full-equivalent squares in the appropriate space below each radiation curve in Table 1.
2) Add R + G + B for each radiation curve in row: **SUM.**
3) Divide R/sum, G/sum, and B/sum and write in appropriate columns for red value, green value, and blue value.
4) Check if their sum is indeed 1 (or 0.99 or 1.01)
5) The Color Value is: red value − blue value. Write in appropriate column. Note: This Color Value is **NOT** the same as the so-called ''Color Index.''
6) To fill the space for: Color seen, use Table 2 below.

Color Seen	Range of Color Values: Red Value − Blue Value
reddish	larger than 0.6
orange	between 0.2 and 0.6
yellowish	between 0.0 and 0.2
white	between −0.1 and 0.0
bluish-white	smaller than −0.1

B) THE LUMINOSITY OF STARS

As defined above, by **LUMINOSITY** we understand the total energy of radiation emitted per second over all wavelength ranges. Depending on the temperature of the emitter, all or part of the electromagnetic spectrum is radiated, from X-rays to radio waves. In this unit, however, we are interested only in the visible part of the spectrum. The human eye is most sensitive to a wavelength of 5,550 Å, in the green wavelength range. As the wavelengths approach the violet range (shorter than about 4,000 Å) as well as the far red range (longer than about 7,000 Å), the sensitivity decreases rapidly. A fair estimate of the ''visible'' luminosity of stars as a function of their temperature can be achieved in the following way:

1) Use question sheet p. 8–23, Table 1. Transfer the value of G (number of squares below the temperature curves in the green range) for each radiation curve to the corresponding place of Table 3.
2) Multiply this G value by the scale factor written in Table 3 below the temperatures. This factor was explained on page 8–18 of this unit. These values are a measure of ''visible'' luminosity. Remember, however, that this luminosity is expressed per unit area of the emitting surface. It would also apply if all stars had the same surface area.

3) The Sun has a surface temperature of about 6,000 °K. We now want to compare the emission of hotter or cooler stars with that of the Sun. Divide all "visible" luminosity values by the luminosity value of the sun, and write in the column: emission relative to sun. The sun's value now is 1.
4) In the next row below, multiply each value by 100. This gives you the percentage emission. The sun's value is 100%.
5) Compare these "visible" luminosity values with those given in Unit 7.2, the H-R diagram. There the actual surface areas of stars are taken into consideration too.

C) LUMINOSITIES AND ABSOLUTE MAGNITUDES

$$\text{The Sun's luminosity is about } 4 \times 10^{26} \left[\frac{\text{joule}}{\text{sec}} = \text{watt.} \right]$$

This value denotes the total energy emitted per second over the whole electromagnetic spectrum. The Sun is, of course, by far the brightest star in the sky, its apparent magnitude m = −26.7. The sun is, however, not at all the most **LUMINOUS** star.

The Sun's distance from us is = 93,000,000 mi = 150,000,000 km = 1 Astronomical Unit (AU) = 4.85×10^{-6} parsec (pc). Imagine the Sun were suddenly transported to a distance of about 30 Light-years (LY) or 10 pc. Naturally, it would then appear very much fainter. Its apparent magnitude m would then be 5.

The absolute magnitude of stars M is a measure of their luminosity. If all stars were at the same distance from us, say 10 pc, then the measure of their brightness m would also be a measure of their luminosity. Or stated another way: The absolute Magnitude M is equivalent to the apparent magnitude m a star would have at a reference distance of 10 pc. The absolute Magnitude of the sun is then M = 5.

Now let us try to assign absolute Magnitude values for the black-body emitters of Table 3. A main sequence star with surface temperature of 4,000 °K or 8,000 °K does not have too different a surface **AREA** from the sun. (See the H-R diagram, Unit 7.2). So we may expect M values from our calculations to be about correct. On the other hand, very cool and very hot stars have quite different surface areas from the sun. We will therefore not attempt to calculate the M values for the 2,000 °K and the 16,000 °K emitters.

1) The sun's emission relative to the sun = 1, of course. This corresponds to M = 5 for the sun. Write this value in the 6,000 °K column.
2) The 4,000 °K emitter has an emission value relative to the sun of

   ________ . From Table 7 of the appendix, this relative emission corres-

   ponds to a magnitude **DIFFERENCE** of about ________ . Since the 4,000 °K emitter radiates less energy than the sun, its absolute Magnitude M therefore has a larger numerical value. Add the magnitude difference to 5 and write in appropriate place. Do the same steps for the 8,000 °K emitter.
3) Consult the finished H-R diagram of Unit 7.2. Read the absolute Magnitude M of a main sequence star with temperature 4,000 °K and write in the column: actual absolute Magnitude M. Do the same steps for the black-body emitter with 8,000 °K.

Unit 8.3

The Doppler Effect

OBJECTIVE

To understand the Doppler effect, and to apply it to a number of astronomical phenomena.

THEORY

You are probably familiar with the Doppler effect in acoustics: Imagine you are standing on the side of a highway and listen to moving cars. As they approach you, you hear a higher frequency of sound; at the moment they pass you, the frequency drops suddenly; and when they recede from you, you hear a still lower frequency.

The same effect occurs with light. For the following discussion, assume a lightsource emitting a single frequency of light, f_0. The frequency f, the wavelength λ, and the speed of light are related by the simple equation

(8.3-1) $$c = f \cdot \lambda$$

Since the speed of light is constant, to every frequency therefore belongs a certain wavelength. To our frequency f_0 belongs a λ_0. The higher the frequency, the shorter the wavelength and vice versa. From now on we will call the wavelength emitted by a lightsource the **LABORATORY** wavelength λ_0.

Case a): **THE LIGHTSOURCE AND THE OBSERVER (YOU) ARE AT REST RELATIVE TO EACH OTHER,** $v_R = 0$. The frequency you see, f, is the same as the frequency emitted f_0. $f = f_0$ if $v_R = 0$. The wavelength seen λ = wavelength emitted λ_0. i.e. **NO DOPPLER SHIFT**.

Case b): **THE LIGHT SOURCE APPROACHES YOU,** or you approach the light source, or the relative speed between the light source and you is one of approach. Then the frequency you see is higher than the emitted frequency. The wavelength you see is shorter than the emitted wavelength. How much the change is, depends on the relative speed V. The higher the speed, the larger the change.

case b: $f > f_0 \quad \lambda < \lambda_0$ if v_R is one of approach

DOPPLER BLUE SHIFT (the wavelength is shifted **TOWARDS** the blue end of the spectrum.)

Case c): **THERE IS A RELATIVE SPEED OF RECESSION BETWEEN THE LIGHT SOURCE AND YOU.** Then the frequency you see is lower than the emitted frequency. The wavelength you see is longer than the emitted wavelength.

case c: $f < f_0 \quad \lambda > \lambda_0$ if v_R is one of recession

DOPPLER RED SHIFT (the wavelength is shifted **TOWARDS** the red end of the spectrum.)

Case d): So far, the direction of relative motion we considered was along the line of sight, either directly towards you or directly away from you. We call this the **"RADIAL"** speed and use the symbol: v_R. **SUPPOSE A LIGHT SOURCE MOVES AT RIGHT ANGLES TO THE LINE OF SIGHT.**

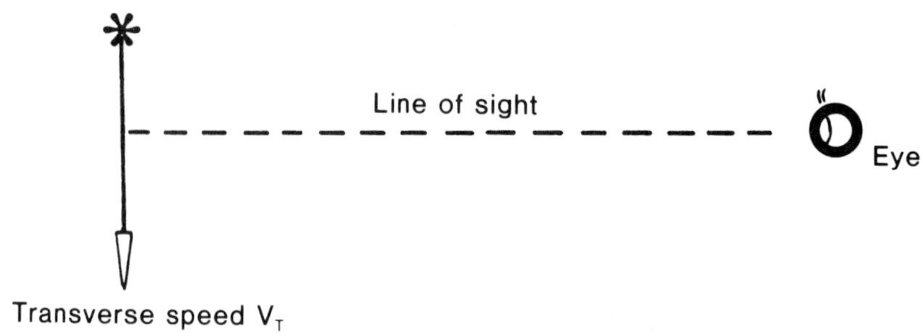

Figure 1

The source neither approaches you nor recedes from you. Its $v_R = 0$. The frequency you see is the same as the frequency emitted. The wavelength you see is the same as the wavelength emitted.

case d: $f = f_0$; $\lambda = \lambda_0$ if speed is transverse

NO DOPPLER SHIFT. (Same as case a).

General case: Obviously, all the cases a) through d) were very special. Consider the general case that a source moves in some arbitrary direction relative to you. What kind of shift do you expect?

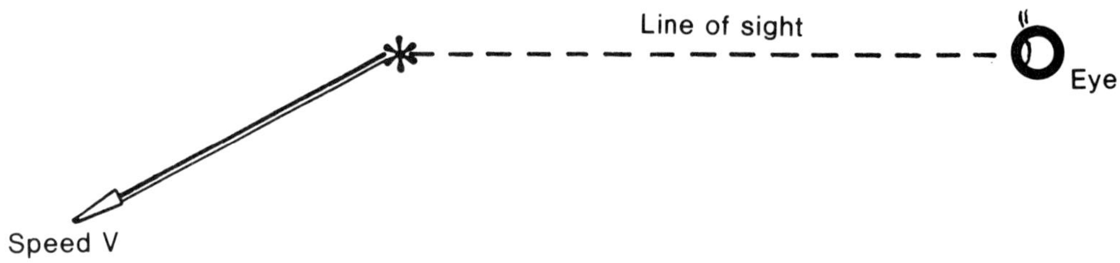

Figure 2

The relative speed in this example is generally one of recession, so we expect a red shift. However, the observed red shift is not as large as if the source would move with speed v directly away from you, along the line of sight. This problem is solved by the resolution of the speed v into its components: One component along the line of sight, v_R, which contributes to the Doppler shift, the other component at right angles to the line of sight, v_T, which does not contribute to the Doppler shift.

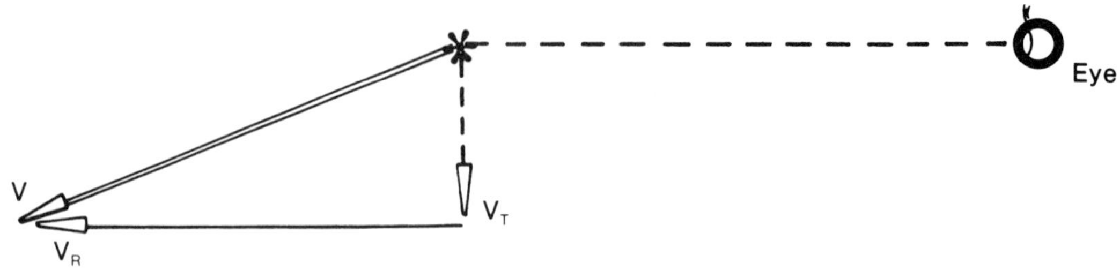

Figure 3

The general expression for the Doppler shift is:

$$f = f_o\left(1 \pm \frac{v_R}{c}\right) \qquad (8.3\text{-}2)$$

the + sign stands for a speed of approach, the − sign for a speed of recession. c is the speed of light. Note: The above equation is valid only as long as the speed of the source is small compared to the speed of light. If we make use of the fact that $f = \frac{c}{\lambda}$, then we can express equation (8.3-2) as

$$\frac{\Delta\lambda}{\lambda_o} = \frac{v_R}{c} \qquad (8.3\text{-}3)$$

where $\Delta\lambda = \lambda - \lambda_o$ in case of red shift
$\Delta\lambda = \lambda_o - \lambda$ in case of blue shift.

Example 1) Suppose a light source emits a single wavelength, e.g. 5,000 Å. This wavelength was measured in the laboratory when source and observer were at rest relative to each other. The wavelength of 5,000 Å looks blue. Now we shoot the light source radially away from us with a speed of 300 km/sec. What will we observe?

From (8.3-3): $\Delta\lambda = \lambda_o \times \frac{v_R}{c}$

$$\Delta\lambda = 5{,}000 \text{ Å} \times \frac{3 \times 10^2 \text{ km/sec}}{3 \times 10^5 \text{ km/sec}} = 5 \text{ Å}.$$

This is a red shift since the source recedes from us. What is the observed wavelength? $\Delta\lambda = \lambda - \lambda_o$; $\lambda = \Delta\lambda + \lambda_o = 5 + 5{,}000 = 5{,}005$ Å. The "shifted" wavelength is **5,005 Å.** Note: Even though we have a red shift, the shifted wavelength looks in no way red. It is merely shifted by 5 Å **TOWARDS** the red (longer) wavelength end of the spectrum.

Example 2) A light source, emitting light with a red wavelength of λ_o = 6,600 Å approaches us at an angle of 45° with a speed of 1,000 km/sec.

a) What kind of Doppler shift will we observe?
b) What is the amount of shift?
c) What is the shifted wavelength?

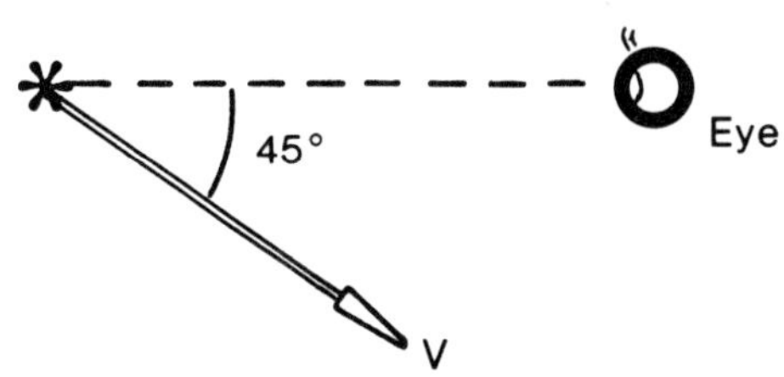

Figure 4

SOLUTION

a) Since the speed is generally one of approach, we will see a blue shift.
b) In order to calculate the amount of shift, we must first resolve the speed v into its components v_R and v_T.

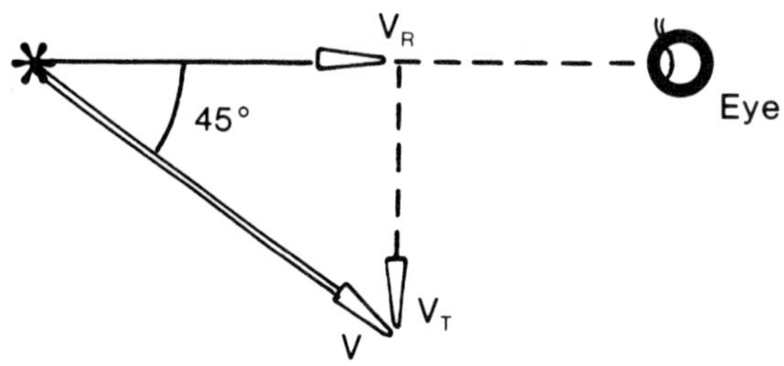

Figure 5

In an isosceles triangle, if the hypothenuse has the length of 1, then each of the other two sides is 0.707 or $(1/\sqrt{2})$ long. In our case, If v = 1,000 km/sec, then both v_R and v_T = 707 km/sec. As before, v_T contributes nothing and the component of v along the line of sight, v_R, will cause the Doppler blue shift:

$$\Delta\lambda = \lambda_0 \times \frac{v_R}{c} = 6{,}600 \times \frac{7.07 \cdot 10^2}{3 \cdot 10^5} = 15.6\ \text{Å}$$

c) The observed wavelength is shifted by 15.6 Å towards the blue end of the spectrum (short wavelength end).

$$\lambda = \lambda_0 - \Delta\lambda = 6{,}600 - 15.6 = 6{,}584.4\ \text{Å}$$

Note again: It is a blue shift, but the shifted wavelength still looks red. Due to the relatively low speed, it is merely shifted slightly towards the blue end of the spectrum.

QUESTIONS Answer the questions in part 1) on p. 8–33.

STAR-LIGHT AND THE DOPPLER EFFECT

Star-light consists of a continuous spectrum with a number of individual spectral lines superimposed. These spectral lines are generally dark, the so-called Fraunhofer lines. They are due to absorption within the topmost layers of stars. If a star approaches us, then **ALL** lines are shifted proportionally towards the blue end. What we observe is the **SHIFTED** line. How do we know what is the wavelength of the line emitted? We would need to know this if we want to find out how fast the star approaches. The wavelengths of the lines are carefully measured. Their spacing depends on what chemical element emits the lines as well as other physical characteristics like temperature, pressure, etc. The branch of physics called "spectroscopy" is concerned with this aspect of star-light. We will assume—with very good justification—that the spectroscopists can tell us what wavelengths are emitted by the star—after identification of the spectral lines. From now on we will assume that a) the emitted wavelengths λ_0 are known, b) the observed wavelengths λ can be measured. We must make some more simplifications: As you have calculated in the questions of part 1), the motion of the earth around the sun contributes a Doppler shift. So does the rotation of the earth around its axis. The motion of the sun through the stars certainly cannot be neglected. However, all these motions are well known and can always be taken into consideration. So, for all the following considerations we will assume that the sun and earth are fixed, i.e. **ALL MOTIONS OF STARS AND GALAXIES WILL BE CONSIDERED RELATIVE TO A STA-**

TIONARY EARTH. The last simplification concerns the actual spectra. Instead of using many spectral lines, we will concentrate on **ONE** or **TWO.**

EXAMPLE

The spectrum of a star is drawn as the top spectrum in Fig. 6, a reference spectrum with the identified laboratory lines directly below. The spectrum as it will be presented for further discussion as well as in the questions you will have to answer is drawn in Fig. 7.

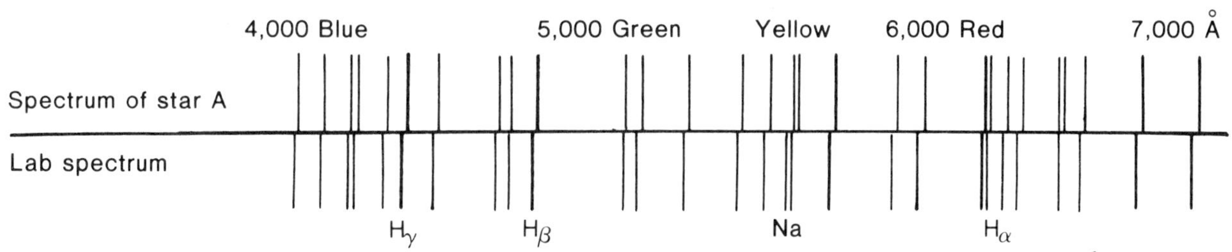

Figure 6

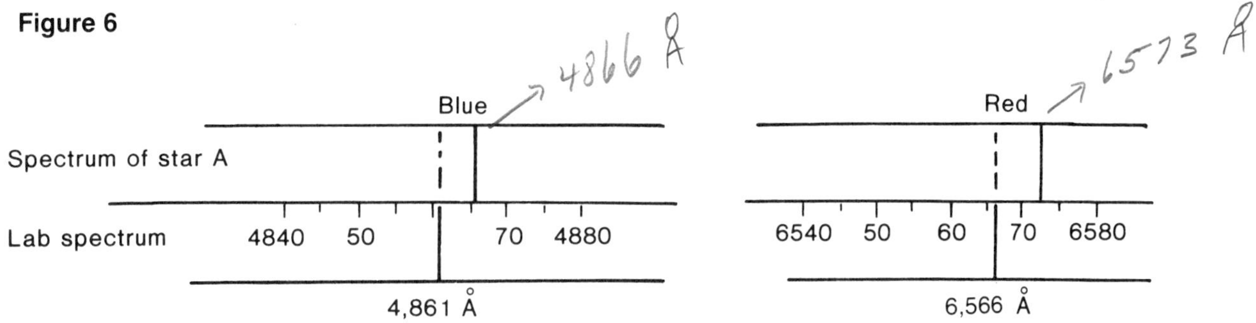

Figure 7

QUESTIONS

Answer the questions, part 2) on p. 8–34.

Consider the following case: A bright star and a black hole revolve around their common center of mass. **IF THEIR MASSES ARE THE SAME, THE CENTER OF MASS IS HALFWAY BETWEEN THEM.** This center shall **NOT** move relative to the sun, in other words, the sun and this "double star" system move together through space. We will treat this problem as if we already know the answer, in order to see which role the Doppler effect plays. See Fig. 8 on the following page:

The observer, behind the spectroscope, is in the same plane as the plane of revolution of the double star system, the plane of this paper. Starting with positions [1] for the star and (1) for the black hole, successive positions are indicated. Inside the orbit are the star's positions, outside the black hole's position. We will consider a single spectral line. For simplicity's sake assume the identified (laboratory) spectral line has a wavelength λ_0 = 6,000 Å. The orbital speed v of both star and black hole shall be 300 km/sec. This orbital speed is indicated in all positions.

QUALITATIVE DISCUSSION

We do not see the black hole, it does not emit **ANY** light, and obviously contributes nothing to any possible Doppler effect. The star in (1) moves away from us. Its orbital speed v is fully along the line of sight. It is therefore the same as v_R. We expect the largest red shift. In (2) the star moves at

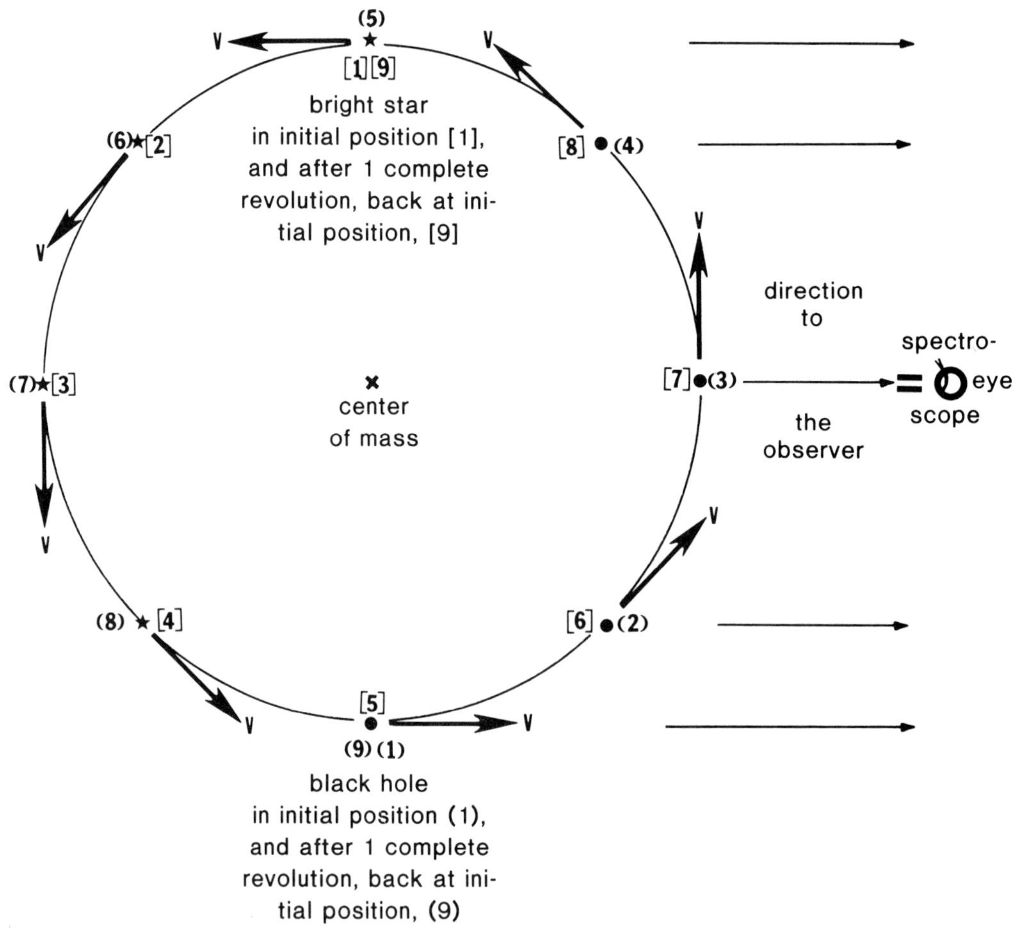

Figure 8

45° away from us. The component of radial speed v_R is 0.707 times the orbital speed v. We expect a smaller red shift than in (1). In position (3) all the motion of the star is transverse. No Doppler shift at all. In position (4), the star approaches us, but not with full orbital speed v. The Doppler shift-effective component v_R is again $0.707 \times v$. In position (5) the star moves with full speed directly towards us. $v_R = v$ and we expect the largest blue shift. What about the Doppler shift in intermediate positions, e.g. between (1) and (2)? The red shift is smaller than in (1) but larger than in (2). Similarly for all other intermediate positions.

QUANTITATIVE ANALYSIS

POSITION (1):

$$\frac{\Delta\lambda}{\lambda_o} = \frac{v_R}{c} = \frac{3 \times 10^2}{3 \times 10^5} = 10^{-3}.\ \Delta\lambda = \lambda_o \times 10^{-3} = 6{,}000 \times 10^{-3}$$

The red shift $\Delta\lambda = 6.0$ Å in position (1) = **MAXIMUM RED SHIFT.**

POSITION (2): $v_R = 0.707\ v = 0.707 \times 300 = 212.1$ km/sec

$$\frac{\Delta\lambda}{\lambda_o} = \frac{v_R}{c} = \frac{2.12 \times 10^2}{3 \times 10^5} = 0.707 \cdot 10^{-3}. \qquad \Delta\lambda = 6{,}000 \times 0.707 \cdot 10^{-3}$$

The **RED** shift $\Delta\lambda = 4.24$ Å in position (2).

POSITION (3): NO DOPPLER SHIFT. $\lambda = \lambda_o = 6{,}000$ Å

POSITION (4): $v_R = 0.707\ v = 212.1$ km/sec

$$\frac{\Delta\lambda}{\lambda_o} = \frac{v_R}{c} = \frac{2.12 \times 10^2}{3 \times 10^5} = 0.707 \cdot 10^{-3} \qquad \Delta\lambda = 6{,}000 \times 0.707 \cdot 10^{-3}$$

The **BLUE** shift $\Delta\lambda = 4.24$ Å in position (4))

POSITION (5): $v_R = v = 300$ km/sec

$$\frac{\Delta\lambda}{\lambda_o} = \frac{v_R}{c} = \frac{3 \times 10^2}{3 \times 10^5} = 10^{-3} \quad \Delta\lambda = 6{,}000 \times 10^{-3}$$

The blue shift $\Delta\lambda = 6.0$ Å in position (5) = **MAXIMUM BLUE SHIFT**. These are all the calculations we need:

Position (6) blueshift, $\Delta\lambda = 4.24$ Å (same as in (4))

Position (7) No shift

Position (8) red shift, $\Delta\lambda = 4.24$ Å (same as in (2))

With this information, we can draw the spectra as seen. In Fig. 9 the (unshifted) laboratory line $\lambda_o = 6{,}000$ Å is drawn on top, and continued by dashed lines through all the other spectra. If we had kept track of the times when we observed all these spectra from (1) through (9), we would also have measured the period of revolution of the stars around their center of mass.

The above described periodic shift of spectral lines reveals a so-called **SINGLE-LINED SPECTROSCOPIC BINARY,** a double star, where only a single component is observed. The other invisible component could be: a) a black hole, b) an extremely cool star, c) any other dark body, e.g. a massive planet.

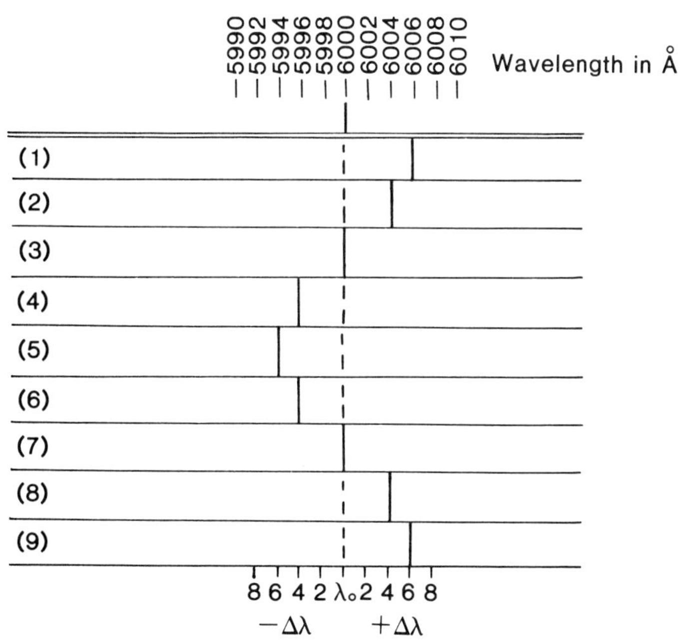

Figure 9

There are many more applications of the Doppler effect in astronomy. One of them, the cosmological Doppler shift, is discussed in detail in Unit 7.4. In the same unit it was mentioned that Cepheid variables are pulsating stars, periodically expanding and contracting. How do we know that? The answer is: by observing a Doppler shift—as is so often the case whenever motion of an object is involved. While the star expands, its matter moves outward and towards the observer—and therefore shows a blue shift. While the star contracts, its matter falls inwards, moving away from the observer and shows a red shift. We would see a periodic shift from blue to red and back to blue, etc. if we continue to look at the star through its cycles of change. How could the periodic shift be distinguished from a single-lined spectroscopic binary? Look again at Figs. 8 and 9.

QUESTION

Answer this question part 3) on p. 8–34. Hint: Consider the Fig. 10 below! Does **ALL** the stellar matter move directly towards you?

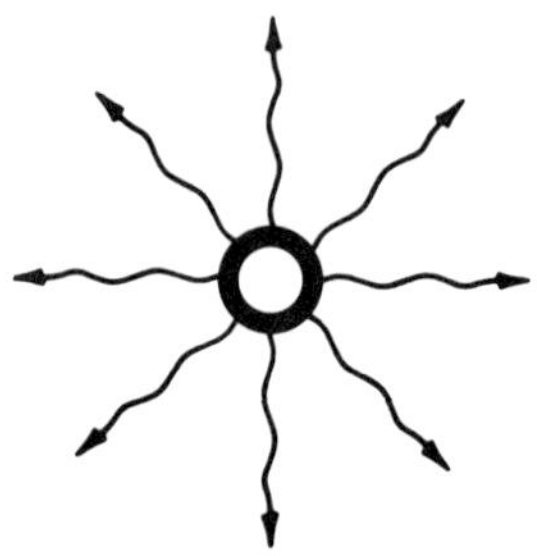

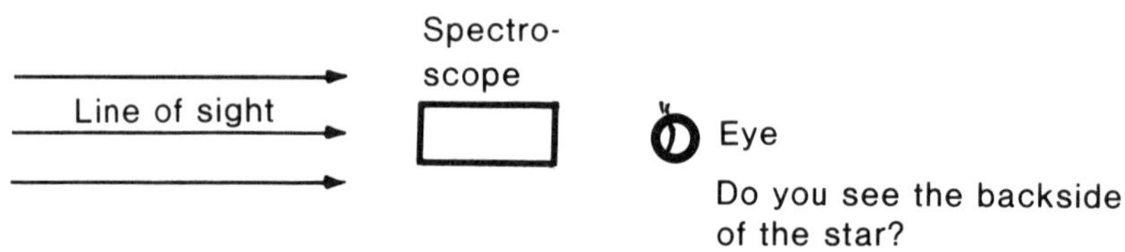

Expanding stellar matter

Figure 10

Another example of the usefulness of the Doppler shift, much closer to home: You have heard of the ‘‘granulation’’ of the solar surface. Hot cells of solar matter stream upward—and then fall back. Looking at such a cell with a spectroscope, we see a blue shift for rising matter, a red shift for falling matter.

QUESTIONS

Answer the questions of part 4) on p. 8–35.

9

APPENDIX

Unit 9.1

Scientific Notation and Significant Figures

OBJECTIVE To become familiar with the power of 10 notation, and to be able to express numbers with meaningful significant figures.

THEORY In astronomy it is necessary to learn to write very large numbers for use with distance, mass, etc. and very small numbers for use with wavelengths of light, mass of elementary particles, etc. A way to write these numbers is the use of scientific notation (sometimes called the power of 10 notation). The following table with the metric prefixes will illustrate the use of this notation. Basically it is nothing more than a short hand way to write large and small numbers by moving the decimal point and multiplying by 10 or 1/10 for each place you move the decimal. As an example consider the distance from the Earth to the Sun as 93,000,000 miles. This is written as 9.3×10^7 miles because the decimal was moved seven places to the left and multiplied by 10 for each place moved. For a small number consider 0.0000314. Each place you move the decimal to the right you must multiply by 1/10 or 10^{-1} (note that the exponent is negative). Thus this number will be 3.14×10^{-5}.

Three arithmetical operations that are used with the numbers are 1) addition-subtraction 2) multiplication 3) division. For these operations you should think of these numbers in two parts, e.g. 3.14×10^{-5} is 3.14 and the 10 to its power. If you are to add the number 2.72×10^{-4} to the number 3.14×10^{-5}, the first thing that must be done is to change the exponent of 10 to the same value for both numbers. To do this you may write 3.14×10^{-5} as 0.314×10^{-4} or you could write 2.72×10^{-4} as 27.2×10^{-5}. Addition now becomes $27.2 \times 10^{-5} + 3.14 \times 10^{-5} = (27.2 + 3.14) \times 10^{-5}$ or $30.34 \times 10^{-5} = 3.034 \times 10^{-4}$. We will return to this number in the section on significant figures.

For multiplication or division it is not necessary to have the exponents of 10 equal. However it is necessary to think of these numbers in their two parts and to learn two rules. For multiplication you would have $(3.14 \times 10^{-5}) \times (2.72 \times 10^{-4})$ as $(3.14)(2.72) \times 10^{-5} \times 10^{-4} = 8.5408 \times 10^{(-5-4)} = 8.5408 \times 10^{-9}$. The rule, multiply the magnitude and add the exponents.

For division the rule is just as simple. $(3.14 \times 10^{-5}) \div (2.72 \times 10^{-4}) = [3.14) \div (2.72)] \times [(10^{-5}) \div (10^{-4})] = 1.1544 \times 10^{-5-(-4)} = 1.1544 \times 10^{-5+4} = 1.1544 \times 10^{-1}$. The rule is, divide the magnitudes and subtract the exponents.

The following table gives you the name, number, power of ten, metric prefix, and symbol for commonly used quantities.

TABLE 1

Name	Number	Power of Ten	Prefix	Symbol
Billion	1,000,000,000	10^9	giga	G
Million	1,000,000	10^6	mega	M
Thousand	1,000	10^3	kilo	K
Hundred	100	10^2	hecto	H
Ten	10	10^1	deca	D
One	1	10^0		
Tenth	.1	10^{-1}	deci	d
Hundredth	.01	10^{-2}	centi	c
Thousandth	.001	10^{-3}	milli	m
Millionth	.000001	10^{-6}	micro	μ
Billionth	.000000001	10^{-9}	nano	n

Unit 9.1 Name ______________________

Date due ____________ Instructor ______________________ Section ____________ Seat ______

Exercises

Express the following numbers in Scientific Notation with one digit to the left of the decimal, e.g. 12,345 = 1.2345×10^4.

1. a) 123 = ______________________

   b) 186,000 = ______________________

   c) 365.25 = ______________________

   d) 0.00123 = ______________________

   e) 5 1/2 million = ______________________

   f) 1 millionth = ______________________

2. $3{,}000{,}000 \times 6{,}000$ = ______________________

3. $3{,}000{,}000 \div 6{,}000$ = ______________________

4. $9.2 \times 10^7 \div 4.6 \times 10^3$ = ______________________

5. $3 \times 10^8 \div 6 \times 10^{-7}$ = ______________________

6. $10{,}000 \div 100$ = ______________________

7. $2{,}000 \times 5{,}000$ = ______________________

8. The speed of light is 186,000 miles/sec, how far will light travel in one day?

   1 light day = ______________________ km.

9. $10^4 + 10^2$ = ______________________

10. 1 billion + 10 million = ____________ billion = ____________ million

11. 1 trillion + 200 billion + 300 million + 400,000 = ____________

12. $10^4 + 2\times10^3 + 3\times10^2 + 4\times10^1 + 5 + 0.6 + 7\times10^{-2} + 8\times10^{-3}$ = ____________

13. $2\times10^{-2} + 10^2 + 20$ = ______________________

14. 10% of 2 trillion = ____________ billion = ____________

SIGNIFICANT FIGURES

Any time a physical measurement is made, the number is an approximation. If you measure the length of a box with a meter stick and record this length as 16.5 cm, your measurement is an approximation. This measurement would indicate that the true length of the box is between 16.45 cm and 16.54 cm. This measurement is to the nearest tenth of a centimeter. If your measurement were to the nearest hundredth of a centimeter you would have recorded 16.50 cm. The first value 16.5 cm indicates three significant figures, the 1, 6 and 5, the second value indicates four significant figures 1, 6, 5 and 0.

Zeros may or may not be significant. If you consider 8700 lbs, this would not necessarily indicate four significant figures. If the measurement were to the nearest 100 lbs the 8 and 7 are significant and the number would be 8.7×10^3 lbs. If measured to the nearest 10 lbs the number would be 8.70×10^3 lbs, and to the nearest pound 8.700×10^3 lbs. Also any zero between two significant digits is itself significant, e.g. 807 lbs.

In mathematical operations your answer would be rounded according to the number of significant figures in the numbers that you started working with. For addition and subtraction the answer will be rounded off to contain digits only as far as the first column containing meaningful digits. For multiplication and division the answer should be rounded off to contain only as many significant figures as are contained in the least exact factor. If you consider the numbers in the last section, note that 3.14×10^{-5} and 2.72×10^{-4} may be written as 0.0000314 and 0.000272. Note that both have three digits that are significant, but the first has digits for seven columns past the decimal while the second has only six columns. We added these two numbers and obtained the answer 3.034×10^{-4}. Note that this answer 0.0003034 implies seven columns of meaningful digits. According to the above rule it is necessary to round off the answer to contain only three significant figures. The last number is 4, hence it is dropped and the answer would be 3.03×10^{-4}. Had the last number been 5 or more you round off to 3.04×10^{-4}. Also in the last section when we multiplied these numbers we quoted the answer 8.5408×10^{-9}. Again there are too many figures in the answer. You must round off the 8 to get 8.541, round off the 1 to get 8.54, with the final answer to be 8.54×10^{-9}.

Consider the following examples with addition.

Examples:

a)	
51.210	5 significant figures
5.516	4 significant figures
0.441	3 significant figures

Note first that all three numbers have implied a meaningful digit at the one thousandth level. Thus your answer must contain three digits to the right of the decimal, and the answer is 57.167.

b) 21.0
 0.0081
 0.00002
———
21.00812

In this case the 21.0 is three significant figures, and the answer would be 21.0 with the .00812 being too small to show any change in the original 21.0 measurement.

For the case of multiplication and division, consider the following examples.

Example:

a) $21.8/4.6 = 4.739130$

According to the above rule, we should have an answer with only two significant figures, thus starting from the right hand side $4.739130 \cong 4.73913 \cong 4.7391 \cong 4.739 \cong 4.74 \cong 4.7$ (answer to two significant figures.)

b) $21.8 \times 4.6 = 98.9 = 99$ (answer)

Unit 9.1 Name ____________________

Date due __________ Instructor __________ Section __________ Seat ______

Exercises

1. Determine the number of significant figures

a) 12.1 __________

b) 561 __________

c) 2.4 __________

d) 0.3771 __________

e) 0.0421 __________

f) 1.0020 __________

g) 9.6×10^7 __________

h) 1.123×10^8 __________

i) 1020 __________

j) 130,000 __________

2. Add, leaving your final answer with the correct number of significant figures.

0.66	0.0041	4.0
2.0	1.002	0.512
711	0.0068	0.272
______	______	______

3. Multiply, leaving your final answer with the correct number of significant figures.

$3.44 \times 0.2 =$ __________

$82.4 \times 0.074 =$ __________

$3.04 \times 5.211 =$ __________

$109.71 \times 0.520 =$ __________

Unit 9.2

Length—Mass—Time

OBJECTIVE To become familiar with the two most common systems of measurement.

THEORY In the English system of measurements, the fundamental quantities are: length, **WEIGHT,** and time. Their units are the foot, pound, and second, respectively.

In the metric system of measure one defines the fundamental units of length, **MASS,** and time. These are the meter, kilogram, and second, respectively. In some cases a smaller system of metric quantities are used. These are the centimeter, gram, and second.

The second (of time) for all three systems is the same. The most common and useful quantities of length, mass, and time are the following.

LENGTH (metric)
1 kilometer (km) = 1000 meters = 10^3 m
1 meter (m) = 100 centimeters = 10^2 cm
1 centimeter (cm) = 10 millimeters (mm)
1 Angstrom (Å) = 10^{-10} meters

LENGTH (English)
1 mile = 1760 yard (yd)
1 yard = 3 feet (ft)
1 foot = 12 inches (in)

CONVERSIONS
1 kilometer = 0.6214 mile
1 mile = 1.6093 km
1 inch = 2.5400 cm
1 cm = 0.3937 inch

MASS (metric)
1 metric ton = 1000 kg
1 kg = 1000 g (gram)
1 milligram = .001 g

WEIGHT (English)
1 ton = 2000 lb
1 pound = 16 oz.

CONVERSIONS
1 kg = 2.2046 lb
1 lb = 0.4536 kg
1 oz = 0.02835 kg

TIME (metric-English)
1 Tropical year = 365.2422 mean solar days
1 day = 24 hours = 86,400 sec
1 hour = 60 minutes
1 minute = 60 seconds

Unit 9.2 Name ______________________________

Date due ____________ Instructor ____________________ Section ____________ Seat ______

Exercises (Express in powers of 10)

1) 1.65 km = ____________ m = ____________ cm

2) 1.65 km = ____________ miles = ____________ ft = ____________ inches

3) 6600 Å = ____________ meters = ____________ cm = ____________ km

4) 5 lbs = ____________ kg

5) 20 kg = ____________ lbs = ____________ metric tons

6) 1.4 kg = ____________ grams

7) 50,000 sec = ____________ % of a day

8) 2500 cm = ____________ ft

9) 75 cm = ____________ inches

10) 9.8 m/sec^2 = ____________ ft/sec^2

11) 1 ft^2 = ______ in^2 = ______ cm^2

12) 1 acre = ______ ft^2 = ______ mi^2

13) 1 mi^2 = ______ acre = ______ ft^2

14) 1 m^2 = ______ cm^2 = ______ km^2

15) 1 mi^2 = ______ km^2 = ______ m^2 = ______ cm^2

16) 1 ft^3 = ______ in^3 = ______ $yard^3$

17) 1 m^3 = ______ liter = ______ cm^3

18) 1 gallon = ______ liter = ______ cm^3

19) 1 liter = ______ gallon

20) 1 km^3 = ______ m^3 = ______ cm^3

21) Assume that you buy a 1/2 acre lot and build a 2,000 ft^2 house on it.

What percentage of the lot is taken up by the house? ______ %

Unit 9.3

Temperature Scales

OBJECTIVE To become familiar with the relation between the Celsius, Kelvin, and Fahrenheit temperature scales.

THEORY Historically the measurement of temperature in America has used a scale called the Fahrenheit temperature scale. In this temperature scale the freezing point of water is at 32 °F, the boiling point of water is at 212 °F. Thus there are 180 Fahrenheit degrees between these points.

Almost all of the rest of the world uses a temperature scale called Celsius, C. On this scale water freezes at 0 °C and boils at 100 °C. Thus on this scale there are 100 degrees between these points. We see that the Fahrenheit degree is smaller than the Celsius degree, i.e. 100/180 = 5/9.

A third scale used for scientific measurement is the absolute or Kelvin temperature scale. On this the minimum temperature, theoretically possible for any material, is chosen as "absolute zero" or 0 °K. The size of the Kelvin degree is the same as that of the Celsius scale, but the zero point has been shifted downward approximately 273 degrees. That is to say that 0 °K = −273 °C. Thus, the ice point of water on the Kelvin scale will be 273 degrees and the boiling point will be 373 degrees K.

The following table will illustrate the relative temperatures of these three scales.

To change from one temperature scale to another the following relations may be used:

$$K = C + 273$$
$$C = K - 273$$
$$C = 5/9(F - 32)$$
$$F = 9/5C + 32$$
$$F = 9/5K - 459$$

Humans live in temperatures of approximately 70 °F which is about 20 °C or 293 °K. The stellar temperatures are commonly thousands of degrees on the Kelvin scale. The figure below shows the relationship between the Fahrenheit and Kelvin scales for stellar temperatures.

TEMPERATURE SCALES

Fahrenheit °F	Celsius °C	Kelvin °K	
212	100	373	boiling point of water
	20	293	room temperature
32	0	273	freezing point of water = melting point of ice
	−39		melting point of Mercury
	−183		boiling point of Oxygen
	−196		boiling point of Nitrogen
	−210		melting point of Nitrogen
	−218		melting point of Oxygen
	−252		boiling point of Hydrogen
	−259		melting point of Hydrogen
	−269		boiling point of Helium
−459	−273	0	absolute zero (molecular motion ceases)

Unit 9.3 Name ____________________

Date due ____________ Instructor ____________________ Section ____________ Seat ______

Exercises

1) 50 °F = ______________ °K

2) 50 °F = ______________ °C

3) 25 °C = ______________ °F

4) 25 °C = ______________ °K

5) 100 °K = ______________ °C

6) 100 °K = ______________ °F

Use the graph to estimate the stellar temperatures for the following stars

7) Star 1 T = 3000 °K = ______________ °F

8) Star 2 T = 22,500 °K = ______________ °F

9) Star 3 T = 30,000 °K = ______________ °F

10) Star 4 T = 45,000 °K = ______________ °F

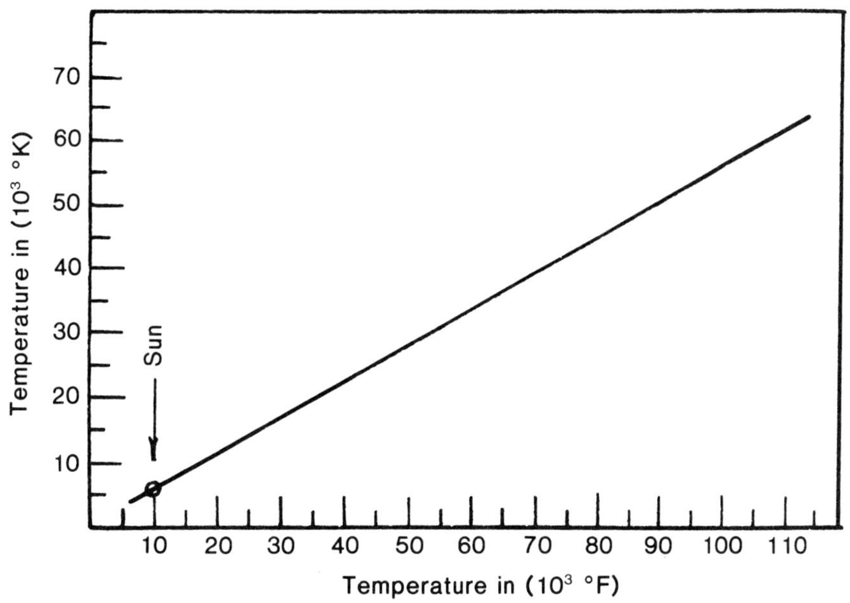

Figure 1

Unit 9.4

Angles and Time

OBJECTIVE To become familiar with angular measures and the use of angles of rotation to measure time.

THEORY The circle is divided into 360 **"DEGREES".** Each degree is subdivided into 60 **"MINUTES OF ARC"** [′]. Each minute of arc is subdivided into 60 **"SECONDS OF ARC"** [″]. A second measure of angle is called the **"RADIAN".** The angle θ expressed in radians [rad] (see p. 7–1) is defined as the ratio of the arc length A to the radius R. If you consider a full circle, the arc length is just its circumference and the angle subtended at the center is 360°. This angle expressed in radians would be

$$\theta = \frac{\text{arc A}}{\text{R}} = \frac{\text{circumference}}{\text{R}} = \frac{2\pi \text{R}}{\text{R}} = 2\pi \text{ radians [rad]}$$

A second example: Consider the angle 45°. This angle subtends one-eighth of the circumference of a circle and hence the arc length would be

$$1/8 \times 2\pi \text{R} = \pi \text{R}/4$$

This angle expressed in radians would be

$$\pi \text{R}/4 \div \text{R} = \pi/4 \text{ [rad]}.$$

We can relate angles of rotation to the quantity time. Each point on the Earth's surface moves in a full circle in 24 hours around the Earth's axis. Hence 24 hours of time correspond to 360° of angle.

Relations between the angular measures in degrees and radians and time measure are given in the following table.

TABLE 1

Angle Degree, Minutes, Seconds	Angle Radian	Time
360°	2π	24 hr
180°	π	12 hr
90°	$\pi/2$	6 hr
57.3°	1.0	3.82 hr
45°	$\pi/4$	3 hr
15°	$\pi/12$	1 hr
1°	1.745×10^{-2}	4 min
1′	2.908×10^{-4}	4 sec
1″	4.847×10^{-6}	0.0667 sec

Unit 9.4 Name ____________________

Date due __________ Instructor ____________________ Section __________ Seat _____

Exercises

1) 6^h15^m = __________ degrees

2) 50 sec of arc = __________ sec

3) 28 deg = __________ time

4) 15° = __________ radians

5) 50 min = __________ degrees

6) 0.5 degree = __________ time

7) Find the number of sec of arc in 360° __________

8) Find the number of min of arc in 360° __________

9) 56 hrs of time = __________ degrees

10) 1.5 days = __________ degrees

TABLE 1. Physical and Astronomical Constants

1. ASTRONOMICAL CONSTANTS

Astronomical unit	$= 1.5 \times 10^{11}$ m
Parsec	$= 2.06 \times 10^{5}$ AU
	$= 3.26$ ly
Light year	$= 9.46 \times 10^{15}$ m
	$= 6.32 \times 10^{4}$ AU
Mass of Earth	$= 6.0 \times 10^{24}$ kg
Mass of Sun	$= 2.0 \times 10^{30}$ kg
Radius of Earth	$= 6.378 \times 10^{6}$ m
Radius of Sun	$= 6.96 \times 10^{8}$ m

2. PHYSICAL CONSTANTS

Gravitational Constant	G	$= 6.67 \times 10^{-11} \dfrac{\text{Newton-m}^2}{\text{kg}^2}$
Speed of light	c	$= 3.0 \times 10^{8}$ m/sec
Boltzmann's Constant	k	$= 1.38 \times 10^{-23}$ joule/°K
Plank's Constant	h	$= 6.62 \times 10^{-34}$ joule-sec
Stefan-Boltzmann Constant	σ	$= 5.67 \times 10^{-8}$ joule/m^2·deg^4·sec
Wien Constant	a	$= 2.9 \times 10^{7}$Å-°K

3. MATHEMATICAL CONSTANTS

$\pi = 3.14159$

$e = 2.71828$

1 radian = 57.3°

TABLE 2. Planetary Data

Planet	Distance from Sun [AU]	Mass (Earth=1)	Diameter (Earth=1)	Rotation Period (sidereal time)	Inclination of Orbit to Ecliptic [°]	Inclination of Equator to Orbit [°]	Sidereal Period [year]	Surface Gravity (Earth=1)	Maximum Apparent magnitudes m
Mercury	0.4	0.055	0.40	58.6 d	7 °	~0°	0.24	0.4	−1.9
Venus	0.7	0.82	0.95	242.9 d	3.4°	~180°	0.62	0.9	−4.4
Earth	1.0	1.00	1.00	23^h56^m	0°	23.5°	1.00	1.0	
Mars	1.5	0.11	0.53	24^h38^m	1.9°	24°	1.88	0.4	−2.8
Jupiter	5.2	317.9	11.20	9^h50^m	1.3°	3°	11.86	2.6	−2.5
Saturn	9.5	95.2	9.49	10^h14^m	2.5°	27°	29.46	1.1	−0.4
Uranus	19.2	14.6	3.69	10^h45^m	0.8°	98°	84.07	1.1	+5.6
Neptune	30.1	17.2	3.50	16^h	0.8°	29°	164.81	1.4	+7.9
Pluto	39.5	~0.1	~0.5	6.4 d	17.2°	?	248.53	—	+14.9

TABLE 3. The Nearest Stars

Name	Spectral Class	M	m	Distance (pc)	Proper Motion	Space Velocity	Radial Velocity	Trans-verse Velocity
Sun	G2	4.9	−26.7					
Proxima Centauri	M5	15.4	11.0	1.3	3.85			15
Alpha Centauri A	G2	4.4	0.0	1.3	3.68	21	−15	14
Alpha Centauri B	K5	5.8	1.4	1.3	3.68	19	−13	14
Barnard's Star	M5	13.2	9.5	1.8	10.30	87	−67	56
Wolf 359	M6	16.6	13.5	2.4	4.84	35	+8	34
Lalande 21185	M2	10.5	7.5	2.5	4.78	64	−53	35
Luyten 726-8 A	M6	15.4	12.5	2.6	3.31	32	+18	26
Luyten 726-8 B	M6	15.9	13.0	2.6	3.31			26
Sirius A	A1	1.5	−1.4	2.7	1.32	11	−5	10
Sirius B	wd	11.4	8.5	2.7	1.32			10
Ross 154	M5	13.3	10.6	2.8	0.67	6	−2	6
Ross 248	M6	14.7	12.2	3.2	1.58	52	−50	14
Epsilon Eridani	K2	6.1	3.7	3.3	0.97	13	+9	9
Ross 128	M5	13.5	11.1	3.4	1.40	16	−8	14
61 Cygni A	K6	7.5	5.2	3.4	5.22	66	−40	52
61 Cygni B	K7	8.3	6.0	3.4	5.22			52
Luyten 789-6	M6	14.5	12.2	3.4	3.27	50	−37	33
Procyon A	F5	2.6	0.3	3.5	1.25	12	−2	12
Procyon B	wd	13.1	10.8	3.5	1.25			12
Epsilon Indi	K3	7.0	4.7	3.5	4.67	54	−25	48
Cincinnati 2456 A	M4	11.1	8.9	3.6	2.29	24	+1	24
Cincinnati 2456 B	M5	11.9	9.7	3.6	2.29	26	+9	24
Groombridge 34 A	M2	10.2	8.0	3.6	2.91	31	+9	30
Groombridge 34 B	M4	13.2	11.0	3.6	2.91	33	+13	30
Tau Ceti	G8	5.7	3.5	3.6	1.92	22	−10	20
Lacaille 9352	M2	9.6	7.4	3.7	6.87	73	+6	73
Luyten's Star	M4	12.2	10.1	3.8	3.73	45	+16	42
Lacaille 8760	M1	8.8	6.8	3.9	3.46	42	+14	40
Kruger 60 A	M4	11.8	9.8	4.0	0.87	18	−15	10
Kruger 60 B	M6	13.4	11.4	4.0	0.87	20	−17	10
Kapteyn's Star	M0	11.2	9.2	4.0	8.79	184	+150	106
Ross 614 A	M6	13.3	11.3	4.0	0.97	19	+15	11
Ross 614 B	—	16.8	14.8	4.0	0.97			11

M = Absolute Magnitude, m = apparent magnitude, all velocities in miles/sec. Proper motion in ″/year. A negative radial velocity is one of approach, a positive one of recession.

Note: All Listed stars are Main-Sequence-Stars with the exception of Sirius B, Procyon A and B.

TABLE 4. The Brightest Stars

Star	Name	Spectral Class	MK Class	M	m	C.I.	Distance (pc)	Proper Motion	Space Velocity	Radial Velocity	Transverse Velocity
Sun		G2	V	4.9	−26.7	0.6	4.85×10^{-6}				
α CMa	Sirius	A1	V	1.5	−1.4	0.0	2.7	1.32	19	−8	17
α Car	Canopus	F0	I-II	−3.1	−0.7	0.2	30	0.03		+21	
α Cen		G2	V	4.4	−0.3	0.7	1.3	3.68	34	−25	23
α Boo	Arcturus	K2	III	−0.3	−0.1	1.2	11.0	2.28	120	−5	120
α Lyr	Vega	A0	V	0.5	0.0	0.0	8.1	0.35	19	−14	13
α Aur	Capella	G2	III	−0.6	0.1	0.8	14.0	0.44	42	+30	29
β Ori	Rigel	B8	I	−7	0.1 ±0.1	0.0		0.00		+21	
α CMi	Procyon	F5	IV-V	2.6	0.3	0.4	3.5	1.25	21	−3	21
α Ori	Betelgeuze	M2	I	−5.5	0.4 ±0.3	1.9		0.03		+21	
α Eri	Achernar	B5	V	−2.0	0.5	−0.2	40.0	0.10		+19	
β Cen		B1	III	−3.0	0.6	−0.2	90	0.04		−12	
α Aql	Altair	A7	IV-V	2.3	0.8	0.2	5.1	0.66	31	−26	16
α Cru		B1	IV		0.8	−0.2		0.04		−11	
α Tau	Aldebaran	K5	III	−0.7	0.9 ±0.1	1.5	21.0	0.20	58	+54	20
α Sco	Antares	M1	I	−4.5	0.9 ±0.1	1.8		0.03		−3	
α Vir	Spica	B1	V	−3.6	0.9 ±0.1	−0.2	80	0.05		+1	
β Gem	Pollux	K0	III	1.0	1.2	1.0	11.0	0.63	32	+3	32
α PsA	Fomalhaut	A3	V	2.0	1.2	0.1	6.9	0.37	14	+7	12
α Cyg	Deneb	A2	I	−7	1.3	0.1		0.00		−5	
β Cru		B0	III	−4.6	1.3	−0.2		0.05		+20	
α Leo	Regulus	B7	V	−0.7	1.4	−0.1	26.0	0.25	30	+4	30
ϵ CMa	Adhara	B2	II		1.5	−0.2		0.00		+27	
α Gem	Castor	A1	V	0.9	1.6	0.0	14.0	0.20	14	+6	13
λ Sco	Shaula	B1	V		1.6	−0.2		0.03		0	
γ Ori	Bellatrix	B2	III	−4.2	1.6	−0.2		0.02		+18	
β Tau	Elnath	B7	III	−3.2	1.7	−0.1		0.18		+8	

M = Absolute Magnitude, m = apparent magnitude, C.I. = Color Index. All velocities in km/sec. Proper Motion in ″/year. A negative radial velocity is one of approach, a positive one of recession.

Between 1771 and 1786 a comet hunter Charles Messier and his colleague Pierre Mechain published lists of 109 objects that could be mistaken for comets. From this M 40, M 73, M 91 and M 102 have been omitted. M 40 is a double star, M 73 has only 4 stars, M 91 has not been observed by others, and M 102 is probably a repetition of M 101.

Many of the other 105 objects may be observed with binoculars, most can be seen with a 2½ to 3 inch telescope, and some amateurs report finding every object in the catalogue with a 6-inch telescope.

TABLE 5. The Messier Catalogue

M	NGC	R.A.	DEC.	mag	Constellation	Discription	Date Observed
1	1952	05h31m	+22°	9	Taurus	Crab Nebula	________
2	7089	21h31m	−1°	7	Aquarius	Glob. Cluster	________
3	5272	13h40m	+28°	6	Canes Venatici	Glob. Cluster	________
4	6121	16h22m	−26°	6	Scorpius	Glob. Cluster	________
5	5904	15h17m	+2°	6	Serpens	Glob. Cluster	________
6	6405	17h39m	−32°	6	Scorpius	Open Cluster	________
7	6475	17h53m	−35°	5	Scorpius	Open Cluster	________
8	6523	18h 2m	−24°		Sagittarius	Nebula	________
9	6333	17h18m	−19°	8	Ophiuchus	Glob. Cluster	________
10	6254	16h56m	−4°	6	Ophiuchus	Glob. Cluster	________
11	6705	18h50m	−6°	7	Scutum	Open Cluster	________
12	6218	16h46m	−2°	7	Ophiuchus	Glob. Cluster	________
13	6205	16h41m	+37°	6	Hercules	Glob. Cluster	________
14	6402	17h37m	−3°	8	Ophiuchus	Glob. Cluster	________
15	7078	21h29m	+12°	6	Pegasus	Glob. Cluster	________
16	6611	18h18m	−14°	7	Serpens	Open Cluster	________
17	6618	18h20m	−16°	7	Sagittarius	Nebula	________
18	6613	18h19m	−17°	7	Sagittarius	Open Cluster	________
19	6273	17h01m	−26°	7	Ophiuchus	Glob. Cluster	________
20	6514	18h01m	−23°		Sagittarius	Nebula	________
21	6531	18h03m	−23°	7	Sagittarius	Open Cluster	________
22	6656	18h35m	−24°	5	Sagittarius	Glob. Cluster	________
23	6494	17h56m	−19°	6	Sagittarius	Open Cluster	________
24	6603	18h17m	−18°	6	Sagittarius	Open Cluster	________
25	IC 4725	18h31m	−19°	6	Sagittarius	Open Cluster	________
26	6694	18h44m	−9°	9	Scutum	Open Cluster	________
27	6853	19h59m	+23°	8	Vulpecula	Nebula	________
28	6626	18h23m	−25°	7	Sagittarius	Glob. Cluster	________
29	6913	20h23m	+38°	8	Cygnus	Open Cluster	________
30	7099	21h39m	−23°	8	Capricornus	Glob. Cluster	________
31	224	00h42m	+41°	4	Andromeda	Galaxy	________
32	221	00h42m	+41°	9	Andromeda	Galaxy	________
33	598	01h33m	+31°	6	Triangulum	Galaxy	________
34	1039	02h41m	+43°	6	Perseus	Open Cluster	________
35	2168	06h08m	+24°	6	Gemini	Open Cluster	________
36	1960	05h35m	+34°	6	Auriga	Open Cluster	________
37	2099	05h55m	+33°	6	Auriga	Open Cluster	________
38	1912	05h27m	+36°	6	Auriga	Open Cluster	________
39	7092	21h32m	+48°	6	Cygnus	Open Cluster	________
41	2287	06h46m	−21°	6	Canis Major	Open Cluster	________
42	1976	05h34m	−6°		Orion	Nebula	________
43	1982	05h35m	−5°		Orion	Nebula	________

TABLE 5 (Continued)

M	NGC	R.A.	Dec.	mag	Constellation	Description	Date Observed
44	2632	08h40m	+20°	4	Cancer	Open Cluster	
45	—	03h46m	+24°	2	Taurus	Open Cluster	
47	2422	07h36m	−14°	5	Pupis	Open Cluster	
48	2548	08h13m	−6°	6	Hydra	Open Cluster	
49	4472	12h29m	+8°	9	Virgo	Galaxy	
50	2323	07h02m	−8°	7	Monoceros	Open Cluster	
51	5194	13h29m	+47°	8	Canes Venatici	Galaxy	
52	7654	23h23m	+61°	7	Cassiopeia	Open Cluster	
53	5024	13h12m	+18°	8	Coma Beren.	Glob. Cluster	
54	6715	18h54m	−31°	8	Sagittarius	Glob. Cluster	
55	6909	19h39m	−31°	6	Sagittarius	Glob. Cluster	
56	6779	19h16m	+30°	8	Lyra	Glob. Cluster	
57	6720	18h53m	+33°	9	Lyra	Nebula	
58	4579	12h37m	+12°	10	Virgo	Galaxy	
59	4621	12h41m	+12°	10	Virgo	Galaxy	
60	4649	12h43m	+12°	9	Virgo	Galaxy	
61	4303	12h21m	+5°	10	Virgo	Galaxy	
62	6266	17h00m	−30°	7	Scorpius	Glob. Cluster	
63	5055	13h15m	+42°	9	Canes Venantici	Galaxy	
64	4826	12h56m	+22°	9	Coma Beren.	Galaxy	
65	3623	11h18m	+13°	10	Leo	Galaxy	
66	3627	11h19m	+13°	9	Leo	Galaxy	
67	2682	08h50m	+12°	7	Cancer	Open Cluster	
68	4590	12h38m	−27°	8	Hydra	Glob. Cluster	
69	6637	18h30m	−32°	8	Sagittarius	Glob. Cluster	
70	6681	18h42m	−32°	8	Sagittarius	Glob. Cluster	
71	6838	19h53m	+19°	7	Sagitta	Glob. Cluster	
72	6981	20h52m	−13°	9	Aquarius	Glob. Cluster	
74	628	01h36m	+16°	10	Pisces	Galaxy	
75	6864	20h05m	−22°	8	Sagittarius	Glob. Cluster	
76	650	01h41m	+51°	11	Perseus	Nebula	
77	1068	02h42m	0°	9	Cetus	Galaxy	
78	2068	05h46m	0°		Orion	Nebula	
79	1904	05h23m	−25°	7	Lepus	Glob. Cluster	
80	6093	16h16m	−23°	7	Scorpius	Glob. Cluster	
81	3031	09h54m	+69°	7	Ursa Major	Galaxy	
82	3034	09h54m	+70°	9	Ursa Major	Galaxy	
83	5236	13h36m	−30°	8	Hydra	Galaxy	
84	4374	12h24m	+13°	10	Virgo	Galaxy	
85	4382	12h24m	+18°	10	Coma Beren.	Galaxy	
86	4406	12h25m	+13°	10	Virgo	Galaxy	
87	4486	12h30m	+13°	9	Virgo	Galaxy	
88	4501	12h31m	+15°	10	Coma Beren.	Galaxy	
89	4552	12h35m	+13°	10	Virgo	Galaxy	
90	4569	12h36m	+13°	10	Virgo	Galaxy	
92	6341	17h17m	+43°	6	Hercules	Glob. Cluster	
93	2447	07h44m	−24°	6	Puppis	Open Cluster	
94	4736	12h50m	+41°	8	Canes Venatici	Galaxy	
95	3351	10h43m	+12°	10	Leo	Galaxy	
96	3368	10h46m	+12°	10	Leo	Galaxy	
97	3587	11h14m	+55°	11	Ursa Major	Nebula	

TABLE 5 (Continued)

M	NGC	R.A.	DEC.	mag	Constellation	Description	Date Observed
98	4192	12h13m	+15°	10	Coma Beren.	Galaxy	______
99	4254	12h18m	+15°	10	Coma Beren.	Galaxy	______
100	4321	12h22m	+16°	10	Coma Beren.	Galaxy	______
101	5457	14h03m	+54°	8	Ursa Major	Galaxy	______
103	581	01h32m	+61°	7	Cassiopeia	Open Cluster	______
104	4594	12h39m	−12°	8	Virgo	Nebula	______
105	3379	10h47m	+13°	10	Leo	Galaxy	______
106	4258	12h18m	+47°	9	Canes Venatici	Galaxy	______
107	6171	16h32m	−13°	9	Ophiuchus	Glob. Cluster	______
108	3556	11h11m	+56°	11	Ursa Major	Galaxy	______
109	3992	11h57m	+54°	11	Ursa Major	Galaxy	______

Glob. Cluster = Globular cluster, all coordinates are approximately the values for 1980, all magnitudes are apparent, and are rounded to the nearest whole number.

NAMED MESSIER OBJECTS

M1	Crab Nebula
M8	Lagoon Nebula
M13	Hercules Cluster
M17	Horseshoe Nebula
M20	Trifid Nebula
M27	Dumbbell Nebula
M31	Andromeda Galaxy
M42	Great Nebula in Orion
M44	Beehive Cluster
M45	Pleiades
M51	Whirlpool Galaxy
M57	Ring Nebula
M97	Owl Nebula
M104	Sombrero Nebula

TABLE 6. The Variable Stars

Type	Spectral Class	Period (Days)	Average Absolute Magnitude (M)	Change in Magnitude (ΔM)	Example
Cepheids (type I)	F to G	3 to 50	−1.5 to −5	0.1 to 2	Delta Cep
Cepheids (type II)	F to G	5 to 30	0 to −3.5	0.1 to 2	W Vir
RR Lyrae	A to F	hours	0 to +1	1 to 2	RR Lyr
Beta Cephei	B	0.1 to 0.3	−2 to −4	0.1	Beta Cep
Delta Scuti	F	hours	0 to 2	0.25	Delta Sct
Mira Type	M	80 to 600	+2 to −2	2.5	Omicron Cet (Mira)
Semiregular	M	30 to 2000	0 to −3	1 to 2	Alpha Ori (Betelgeuze)
Irregular	All	Irregular	near 0	0 to 4	Pi Gru
RV Tauri	G to K	30 to 150	−2 to −3	0 to 3	RV Tau
Spectrum Var	A	1 to 25	0 to +1	0.1	

The following table gives the intensity ratios for some selected stellar magnitude differences. If we identify the brighter star with I_b for its intensity, and the fainter star with I_f for its intensity, the difference in magnitudes is $\Delta m = 2.5\log(I_b/I_f)$ or $\Delta M = 2.5\log(L_b/L_f)$ if the luminosity is used rather than intensity. The first column is the magnitude difference, the second column is the ratio I_b/I_f and the third column is I_f/I_b.

TABLE 7. Stellar Magnitude Scales

Δm or ΔM	I_b/I_f or L_b/L_f	I_f/I_b or L_f/L_b
0.0	1.0	1.0
0.5	1.6	0.63
1.0	2.5	0.40
1.5	4.0	0.25
2.0	6.3	0.16
2.5	10.0	0.10
3.0	16.0	0.063
3.5	25.0	0.04
4.0	40.0	0.025
4.5	63.0	0.016
5.0	100	0.01
6.0	250	0.004
7.0	630	0.0016
8.0	1600	0.00063
9.0	4000	0.00025
10.0	10,000	0.0001
11.0	25,000	0.00004
12.0	63,000	0.000016
15.0	1,000,000	0.000001
20.0	100,000,000	0.00000001

TABLE 8. Coordinates of the Ecliptic (Dec. Has Been Rounded to the Nearest Degree)

R.A. hr.	R.A. Deg.	Dec. Deg.
0	360	0
23	345	−6
22	330	−12
21	315	−17
20	300	−20
19	285	−23
18	270	−23.5
17	255	−23
16	240	−20
15	225	−17
14	210	−12
13	195	−7
12	180	0
11	165	+7
10	150	+12
9	135	+17
8	120	+21
7	105	+22
6	90	+23.5
5	75	+22
4	60	+21
3	45	+17
2	30	+12
1	15	+7

Unit Name ______________________________

Date due ______________ Instructor ______________________ Section ____________ Seat _____

Stellar Magnitudes (apparent); all stars brighter than m = 3 are drawn in

● brighter than m = 1.0

• fainter than m = 1.1 and brighter than m = 2.5

• fainter than m = 2.6 and brighter than m = 4.0

• fainter than m = 4.0

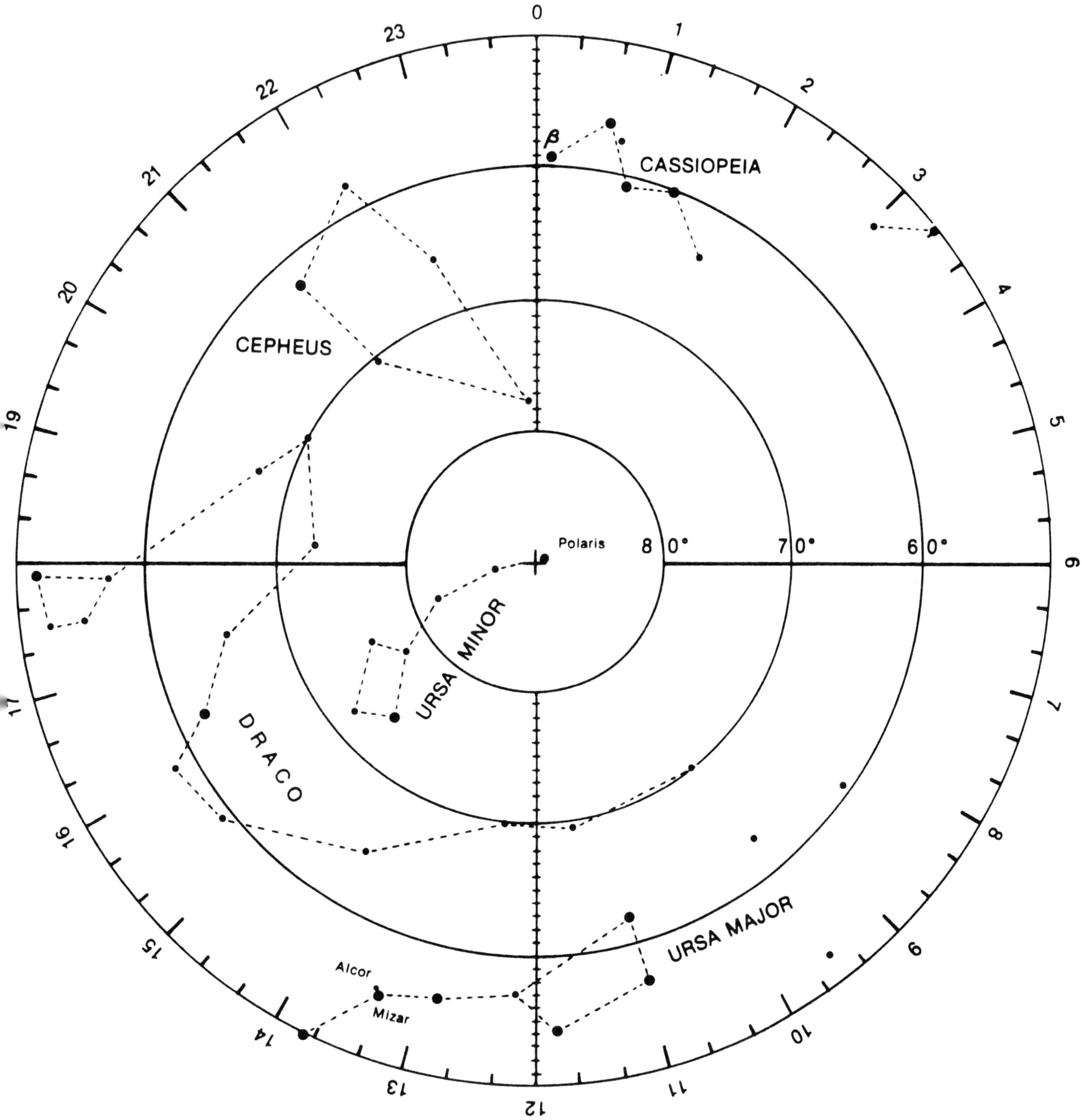

Unit Name ____________________

Date due __________ Instructor __________ Section ______ Seat ____

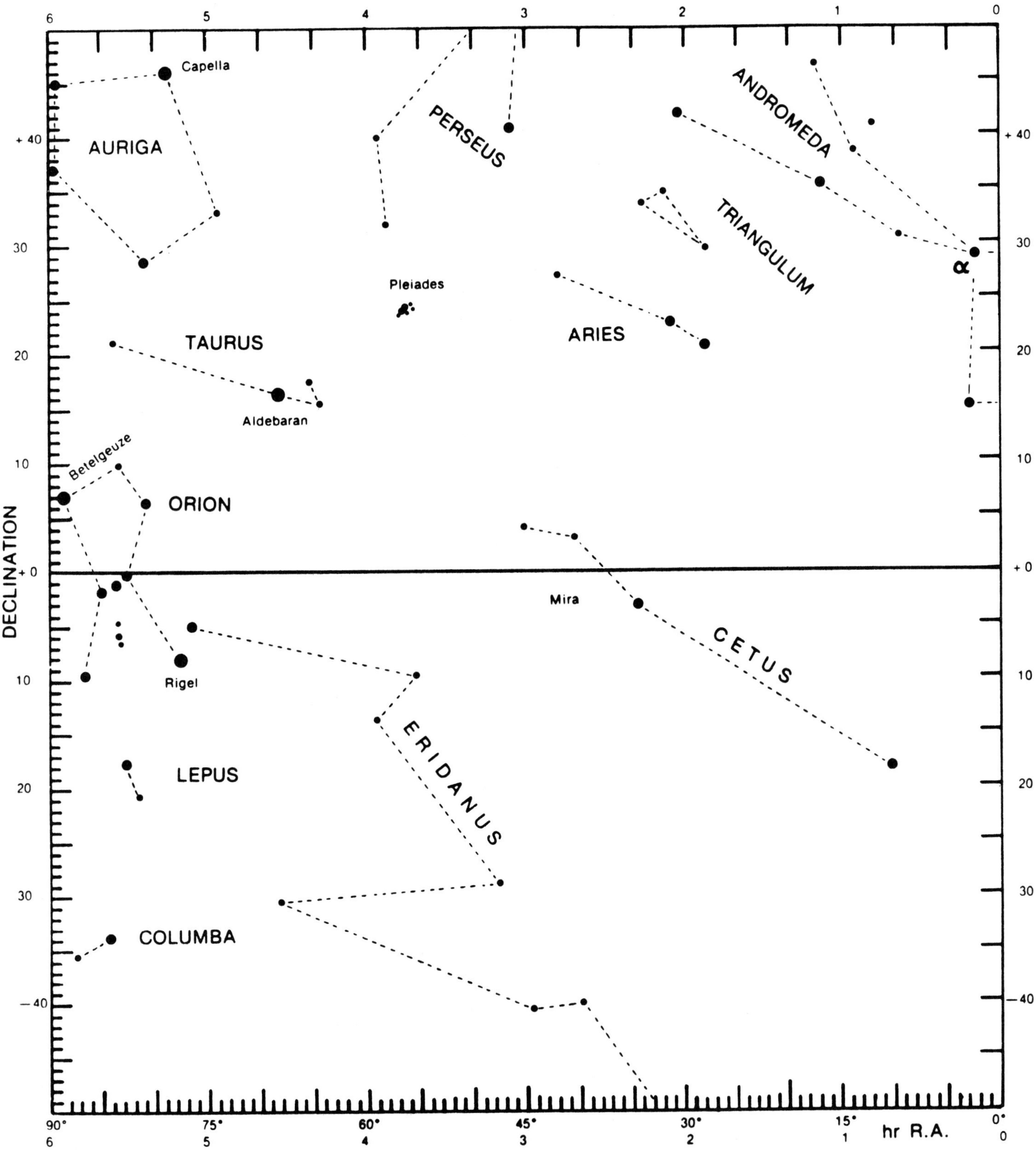

Unit Name ______________________________

Date due ____________ Instructor ____________________ Section __________ Seat _____

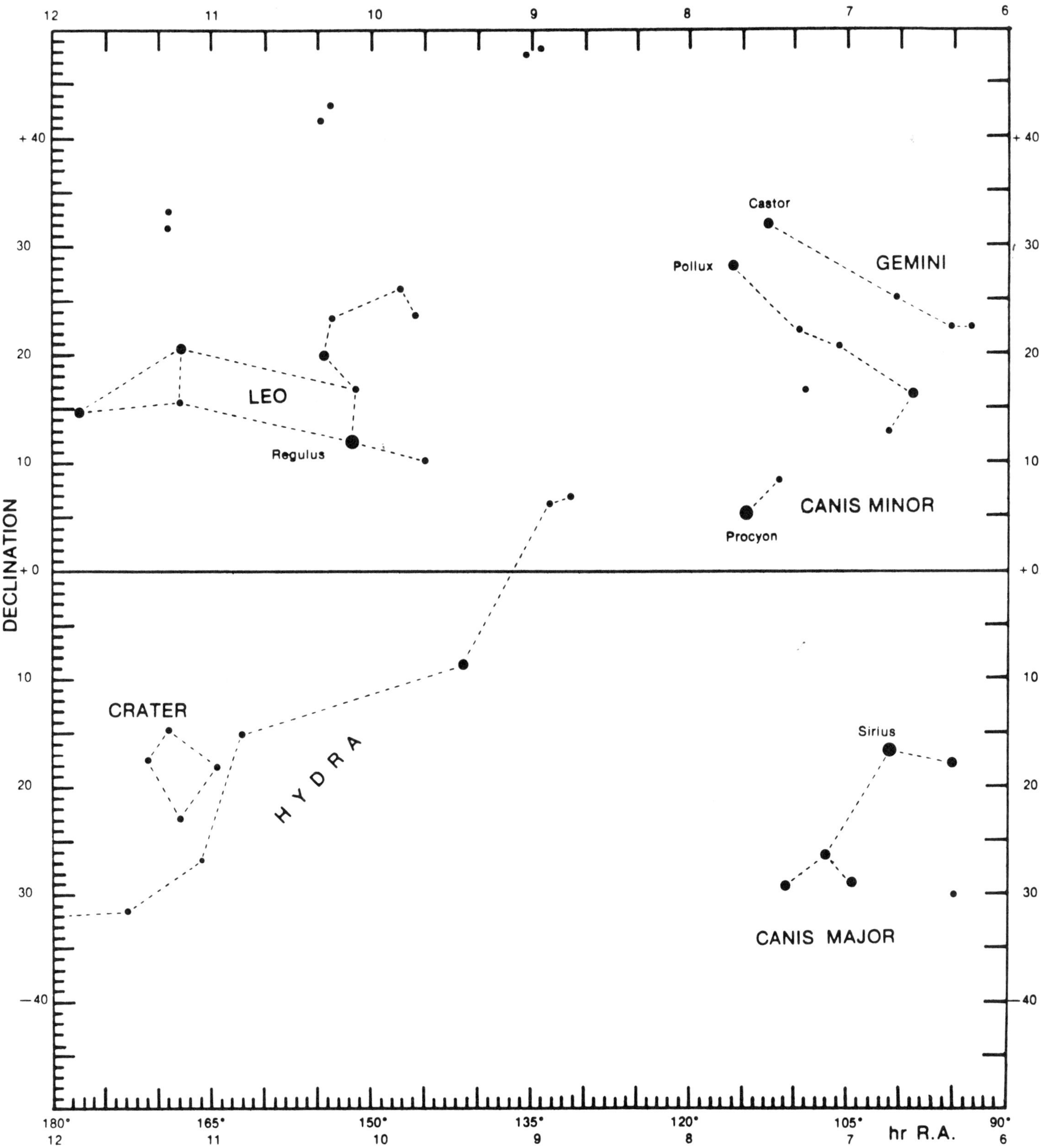

Unit Name ______________________________

Date due ______________ Instructor ______________________ Section ____________ Seat ______

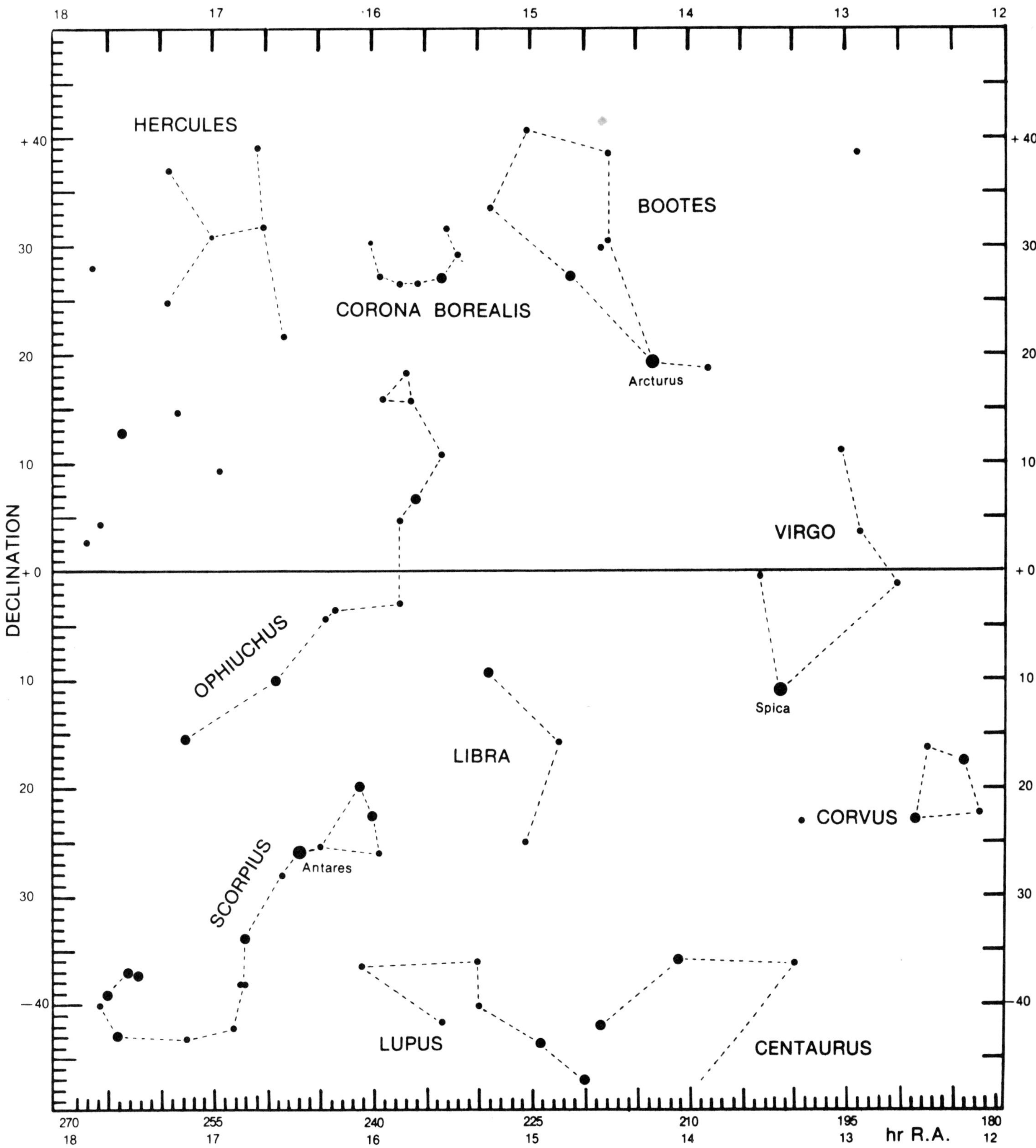

Unit Name ____________________

Date due ____________ Instructor ____________________ Section ________ Seat ____

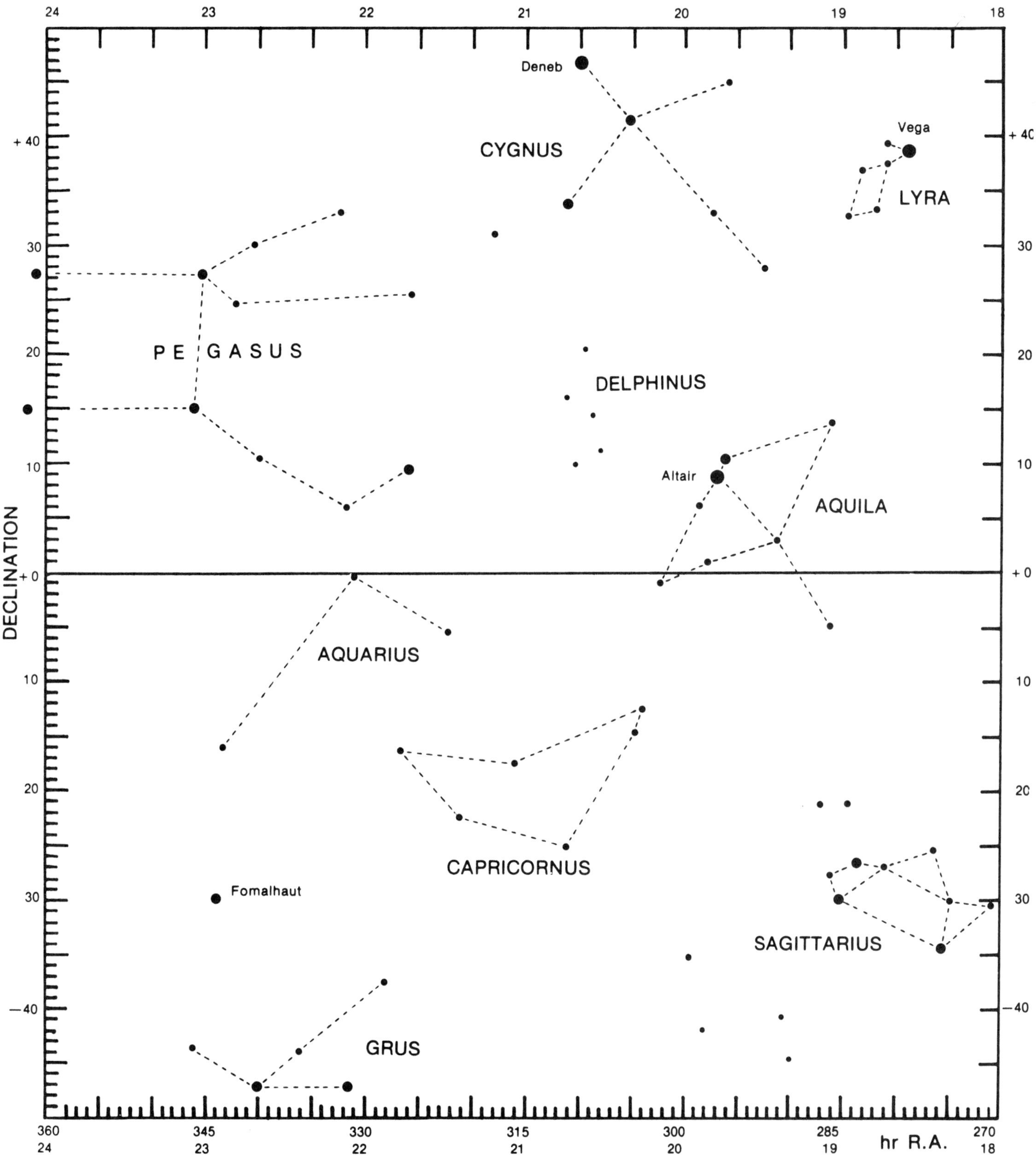

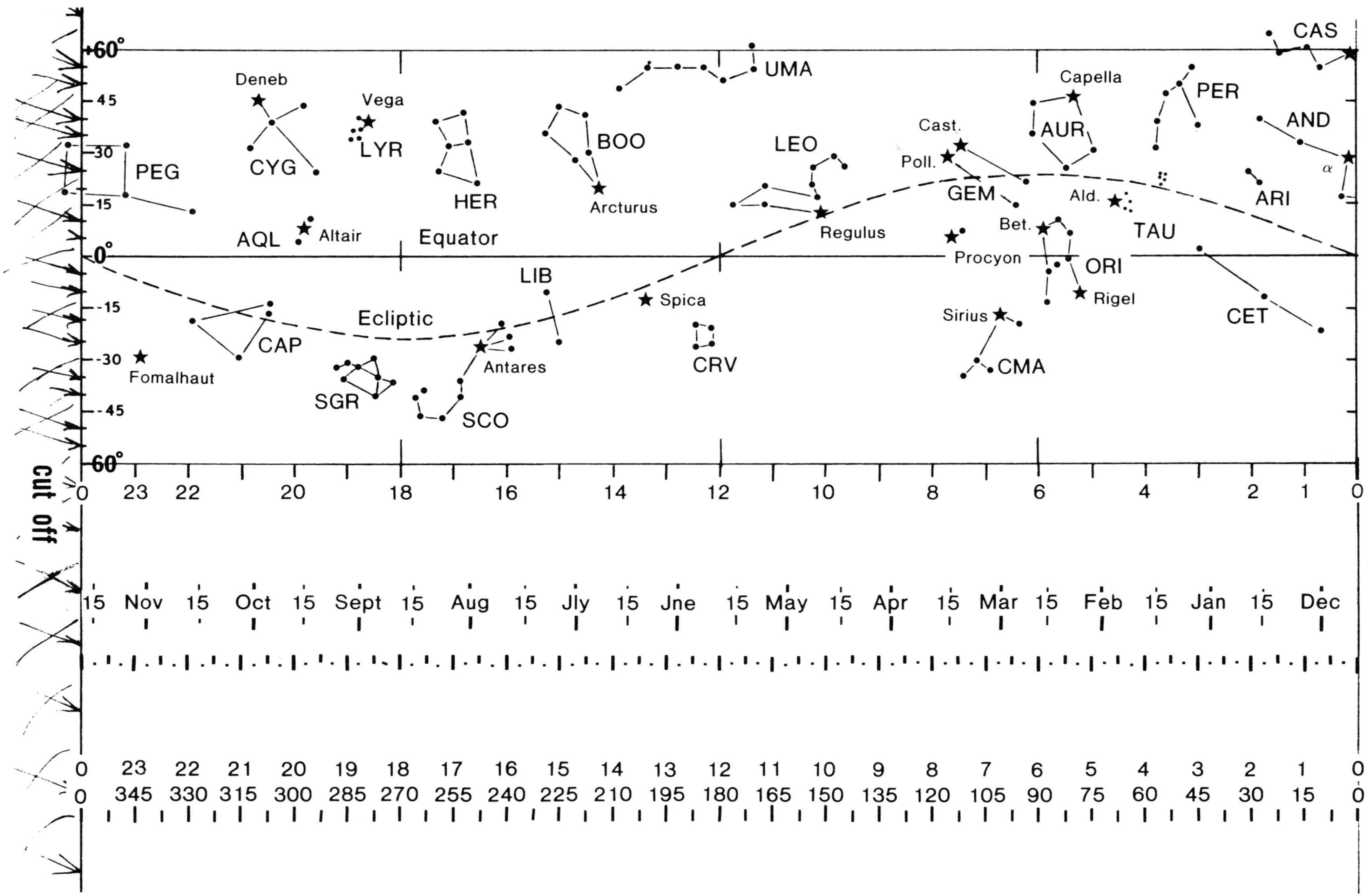

+60°
45
30
15
0°
-15
-30
-45
-60°
CAS
AND
ARI
PER
TAU
Ald.
Capella
AUR
ORI
Rigel
Bet.
Procyon
CET
CMA
Sirius
GEM
Cast.
Poll.
LEO
Regulus
UMA
CRV
Spica
BOO
Arcturus
LIB
Antares
SCO
HER
Equator
Ecliptic
Vega
LYR
SGR
Altair
AQL
Deneb
CYG
CAP
PEG
Fomalhaut
α
0 23 22 20 18 16 14 12 10 8 6 4 2 1 0
cut off
15 Nov 15 Oct 15 Sept 15 Aug 15 Jly 15 Jne 15 May 15 Apr 15 Mar 15 Feb 15 Jan 15 Dec
0 23 22 21 20 19 18 17 16 15 14 13 12 11 10 9 8 7 6 5 4 3 2 1 0
0 345 330 315 300 285 270 255 240 225 210 195 180 165 150 135 120 105 90 75 60 45 30 15 0

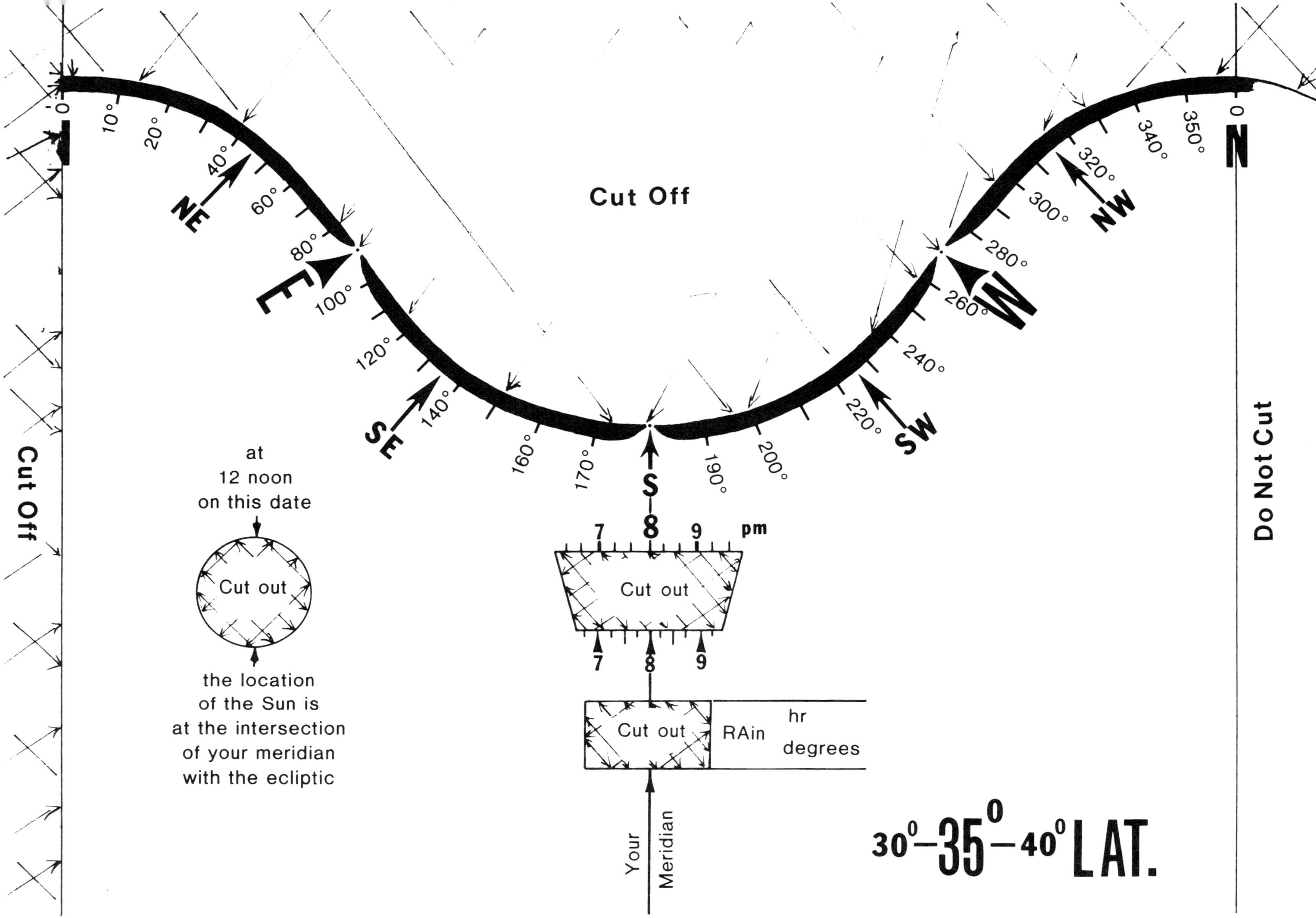
Cut Off
Cut Off
Do Not Cut
0
10°
20°
40°
60°
80°
100°
120°
140°
160°
170°
190°
200°
220°
240°
260°
280°
300°
320°
340°
350°
0
N
NE
E
SE
S
SW
W
NW
N
at
12 noon
on this date
Cut out
the location
of the Sun is
at the intersection
of your meridian
with the ecliptic
7
8
9
pm
Cut out
7
8
9
Cut out
RAin
hr
degrees
Your
Meridian
30°–35°–40° LAT.

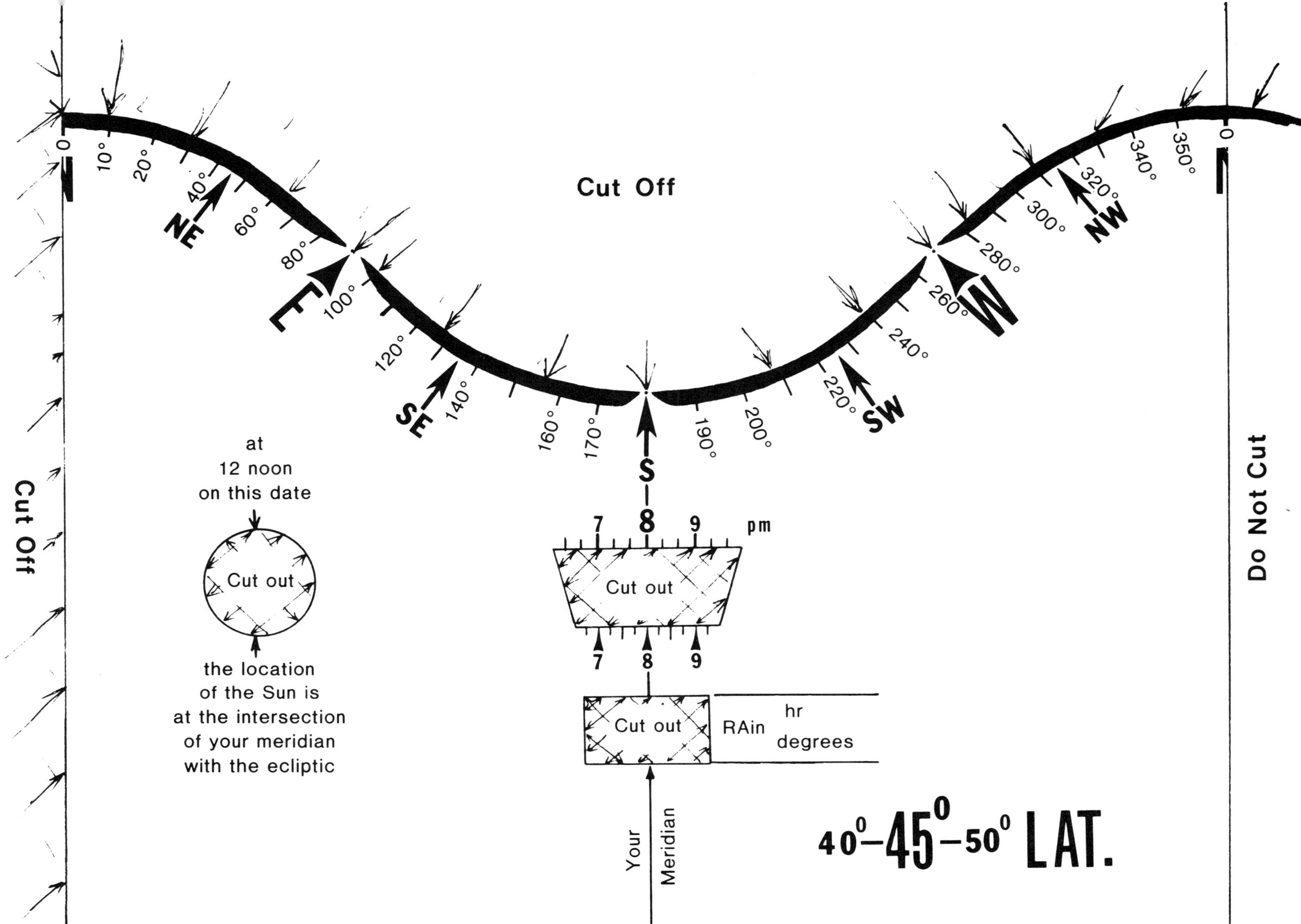
Cut Off
Cut Off
Do Not Cut
0
10°
20°
40°
NE
60°
80°
E
100°
120°
SE
140°
160°
170°
S
190°
200°
220°
SW
240°
260°
W
280°
300°
320°
NW
340°
350°
0
at
12 noon
on this date
Cut out
the location
of the Sun is
at the intersection
of your meridian
with the ecliptic
7
8
9
pm
Cut out
7
8
9
Cut out
RAin
hr
degrees
Your
Meridian
40°–45°–50° LAT.

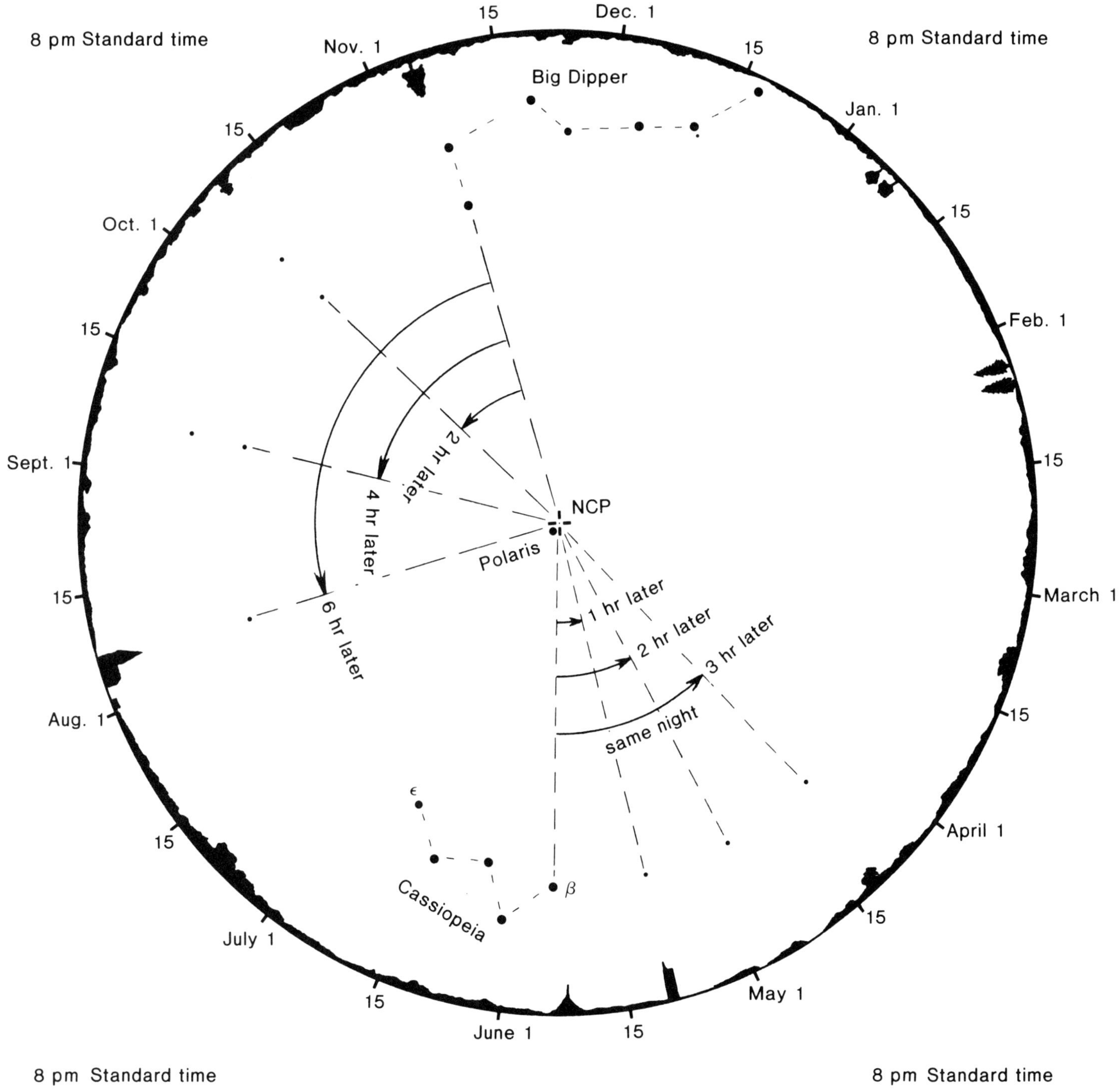
8 pm Standard time
8 pm Standard time
8 pm Standard time
8 pm Standard time
Dec. 1
15
Jan. 1
15
Feb. 1
15
March 1
15
April 1
15
May 1
15
June 1
15
July 1
15
Aug. 1
15
Sept. 1
15
Oct. 1
15
Nov. 1
15
Big Dipper
NCP
Polaris
2 hr later
4 hr later
6 hr later
1 hr later
2 hr later
3 hr later
same night
ϵ
β
Cassiopeia

Unit Name ______________________________

Date due ____________ Instructor ____________________ Section ____________ Seat ______

Ecliptic or Helio-centric coordinates
The orbits of Mercury, Venus, Earth, and Mars are drawn to scale.

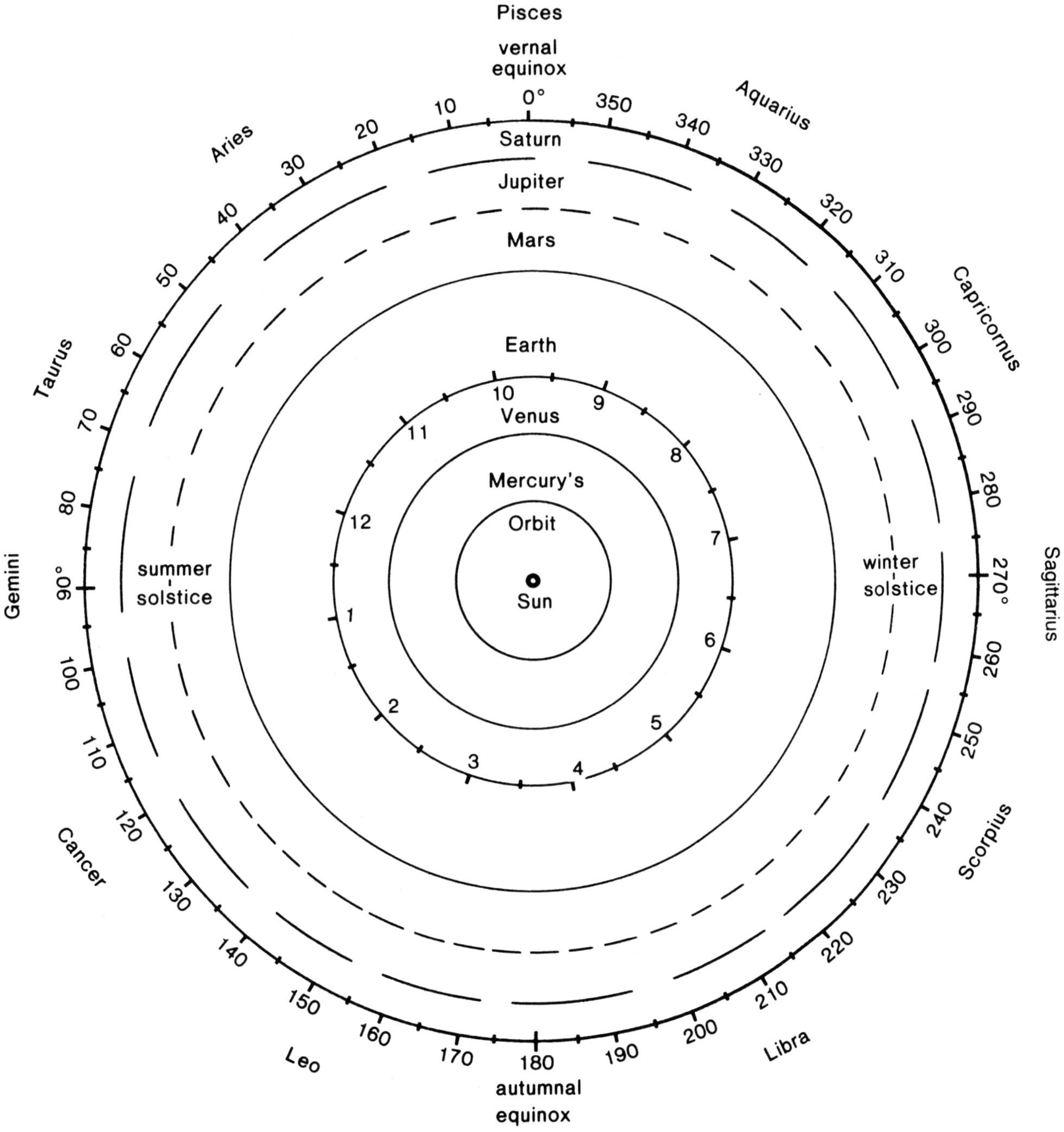

Unit Name ____________________

Date due ____________ Instructor ____________ Section ____________ Seat ______

Ecliptic or Helio-centric Coordinates.
The orbits of Earth, Jupiter, and Saturn are drawn to scale.

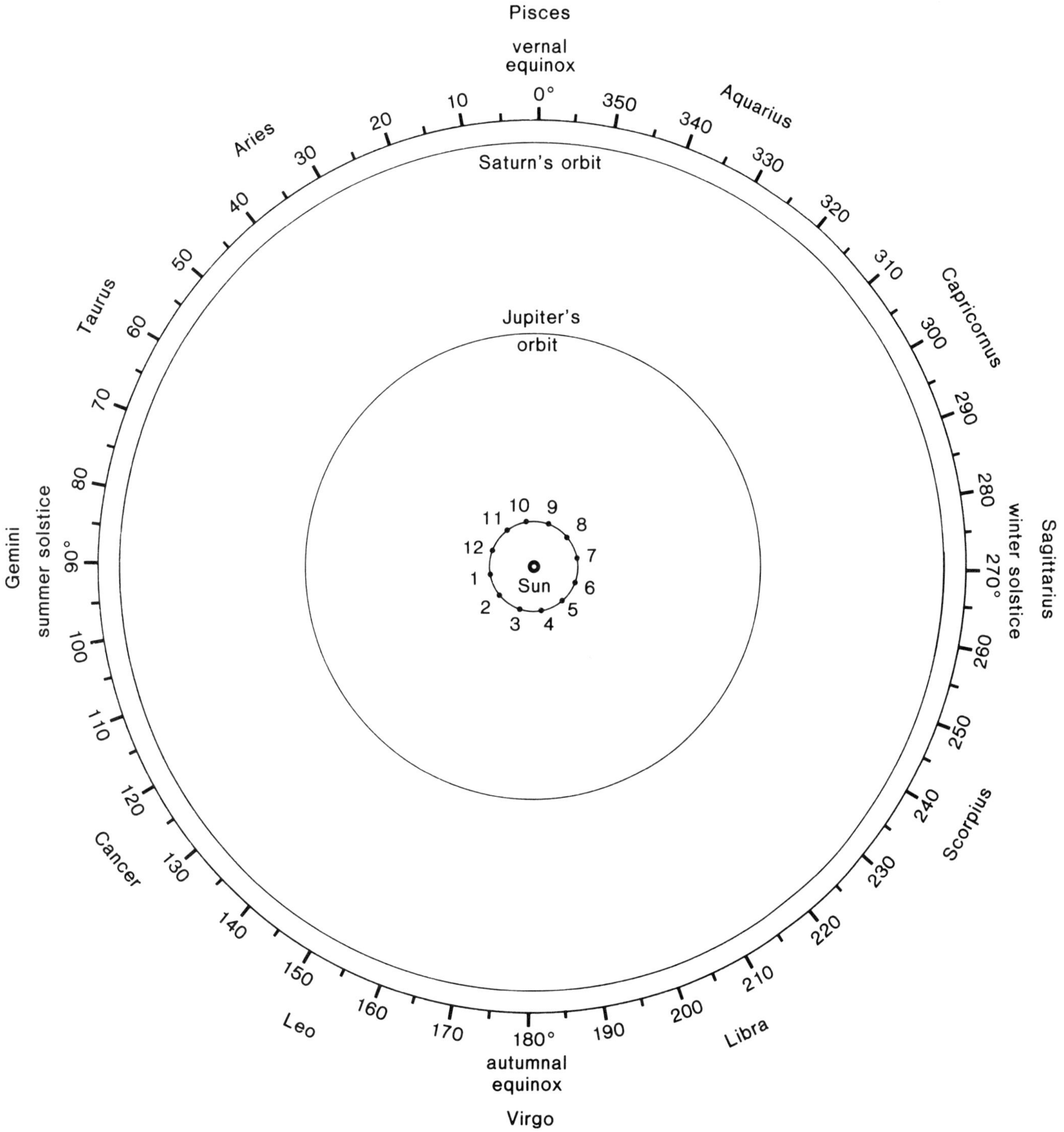

Name ____

Date due ____________ Instructor ____________________ Section __________ Seat ________

	CONSTELLATION Astron. Name	CONSTELLATION Common Name	STAR name(s)	Other Information
1			a)	
			b)	
2			a)	
			b)	
3			a)	
			b)	
4			a)	
			b)	
5			a)	
			b)	
6			a)	
			b)	
7			a)	
			b)	
8			a)	
			b)	
9			a)	
			b)	
10			a)	
			b)	
11				
12				
13				
14				
15				
16				
17				

Name ____

Date due ______________ Instructor ______________________ Section __________ Seat __________

	CONSTELLATION: Astronomical Name Common Name	STAR	COLOR	M–K Class	Other Information
1		a)			
		b)			
2		a)			
		b)			
3		a)			
		b)			
4		a)			
		b)			
5		a)			
		b)			
6		a)			
		b)			
7		a)			
		b)			
8		a)			
		b)			
9		a)			
		b)			
10		a)			
		b)			
11					
12					
13					
14					
15					

STAR MAP FOR THE STAR-TRACKER

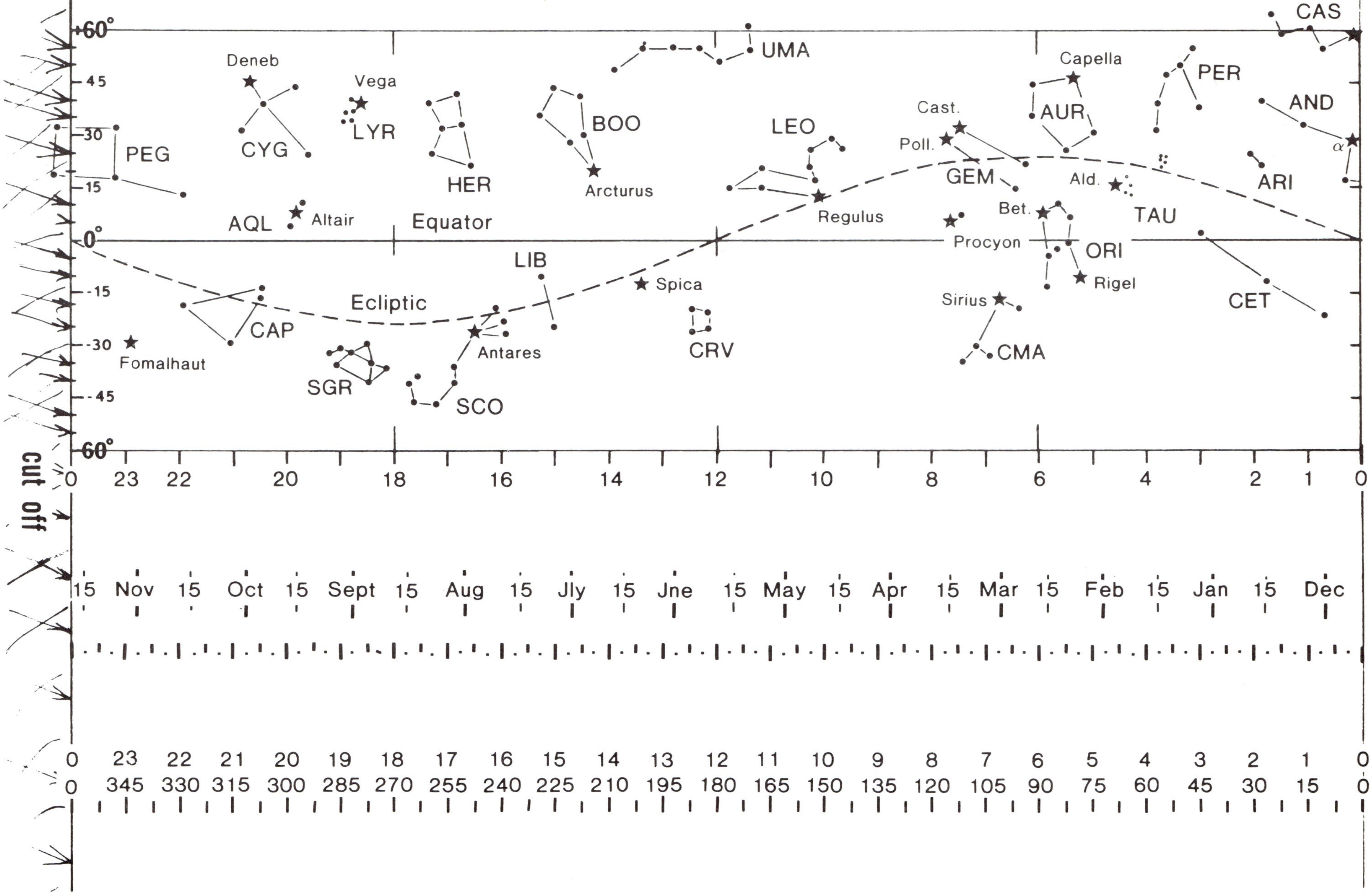